The Invisible Enemy

616.01   CRAWFORD, DOROTHY
CRA   The Invisible
          Enemy

DATE DUE

The Invisible Enemy

A Natural History of Viruses

Dorothy H. Crawford

Department of Medical Microbiology,
The University of Edinburgh Medical School

OXFORD
UNIVERSITY PRESS

OXFORD

UNIVERSITY PRESS

Great Clarendon Street, Oxford OX2 6DP

Oxford University Press is a department of the University of Oxford.
It furthers the University's objective of excellence in research, scholarship,
and education by publishing worldwide in

Oxford New York

Athens Auckland Bangkok Bogotá Buenos Aires
Cape Town Chennai Dar es Salaam Delhi Florence Hong Kong Istanbul
Karachi Kolkata Kuala Lumpur Madrid Melbourne Mexico City Mumbai
Nairobi Paris São Paulo Shanghai Singapore Taipei Tokyo Toronto Warsaw

with associated companies in Berlin Ibadan

Oxford is a registered trade mark of Oxford University Press
in the UK and in certain other countries

Published in the United States by
Oxford University Press Inc., New York

British Library Cataloguing in Publication Data
Data available

Library of Congress Cataloguing in Publication Data
Crawford, Dorothy.
The invisible enemy: a natural history of viruses/Dorothy Crawford.
p. cm.
Includes index.
1. Viruses—Popular works. 2. Medical virology—Popular works. 3. Viruses—Ecology.
I. Title.
QR364.C73 2000   616'.0194—dc21   00-036756
ISBN 0 19 850332 6 (hb)
ISBN 0 19 856481 3 (pbk)

Typeset bt EXPO Holdings, Malaysia
Printed in Great Britain
on acid-free paper by
Clays Ltd, St Ives plc

Foreword

When modern science began to take shape in Europe in the seventeenth century, and even as it developed in the eighteenth century, the concept of a professional scientist was unknown. Those who actively investigated scientific matters, as well as those who were merely interested, were largely well-educated members of the upper classes. Indeed, this situation continued well into the nineteenth century as is evidenced by elections to the Royal Society of London, the UK's national academy of science and the longest established scientific society in the world. During those centuries it was thus usual for well-informed people to be able to understand most of the science, or 'natural knowledge' as it was then called, which was being discovered and discussed at the time.

However, the latter half of the nineteenth century saw advances in science accelerate and the birth of whole new areas of research with their attendant academic disciplines; as a result, the pursuit of science became a career instead of a hobby. The past few decades have witnessed a veritable explosion in knowledge with hundreds of new subjects of ever greater complexity, huge and costly research undertakings, and a vast worldwide army of highly specialized scientists. The ordinary citizen, however generally well educated, has long been left behind in this unimaginably complicated era of new ideas—confusing, impenetrable, and unintelligible, except to experts.

This would not perhaps matter greatly were it not that the technological results of scientific progress impinge more and more on the way society functions and, even more important, on the way the individuals who make up society live their own lives. Current acute anxieties over genetic manipulation of food and human reproduction, over the effects of technological advances on pollution and the environment, and over the influence on behaviour of a globally penetrating mass media, are just a few examples of what people perceive as the arcane and threatening progress of science.

An obvious consequence of the public's bafflement over science is to be seen in the present antiscience backlash evident in Western countries which is not only directed against science and scientists, but also against the multinational companies employing the new technologies which advances in science have provided. It is regrettable that scientists have, on the whole, been reluctant to come forward and explain the nature of whatever highly specialized work they may be doing, and have likewise hesitated, particularly in the face of sustained tabloid hysteria, to stand up and make clear the huge benefits to mankind of recent scientific progress.

But without an understanding of what science is about and how it works, the public is not in a position to evaluate findings, make judgements about their impact on daily life, or appreciate the nature of risk or scientific uncertainty. It is therefore greatly to be welcomed that Professor Dorothy Crawford has now set out in this book a layman's straightforward and authoritative guide to virology, a subject constantly affecting our daily lives in both trivial and life-threatening ways. The extent of confusion and misinformation in this area is exemplified by the continuing inability of even the serious press to distinguish the fundamental differences between viruses and bacteria. Professor Crawford's reliable and readily comprehensible account of the viruses of man is thus more than timely and clearly explains, in an engaging way, a field of science which is too often garbled by the uninformed. Her book will fascinate the curious public and should be obligatory reading for all those journalists who so regularly display their ignorance of this important subject.

Sir Anthony Epstein

Preface

In 1969, the United States Surgeon General, assuming that all severe infections could either be prevented or treated, told Congress that 'we can now close the book on infectious diseases'. He was spectacularly wrong. We are in the middle of one of the largest and most widespread pandemics of a deadly virus the world has ever known. The AIDS virus, HIV, is now the world's biggest killer infection and the commonest cause of death in Africa. And since HIV struck in the early 1980s there have been an unprecedented number of infectious disease outbreaks throughout the world, including the highly lethal Hanta, Ebola, and Lassa fever viruses.

Viruses enter our bodies silently and invisibly and then parasitize our cells in a way that is almost totally beyond our control. Until relatively recently viruses pitted themselves against us, the most sophisticated of beings, and came out on the winning side. A single virus, smallpox, killed at least 300 million people in the twentieth century before it was eradicated in 1980, and at that time measles still killed two and a half million children a year. Now that we have ways of dealing with these old adversaries, new ones are emerging. Will we be able to control them too? Or will they get the better of us? How will the battle between man and virus end?

Epidemics of deadly viruses cause fear and panic which the media capitalize on. But too often what we read is inaccurate and overdramatized, so that the word 'virus' has come to have the most sinister connotations. People have a right to accurate information about microbes that invade their bodies. The phrase 'it's only a virus' is commonly used to explain away coughs and colds, fever or tiredness, but do we really know that a virus is the cause? And even when it is a genuine virus infection, do we know which type of virus and how it comes to cause the symptoms?

This book was written to explore these and many related questions and to try to convey to the general reader the fascination of viruses. They consist of no more than a minute piece of genetic material wrapped in a protein coat, but they can cause chaos. It is difficult not to feel a sneaking admiration for the way these tiny creatures manipulate us so successfully. Inevitably, throughout the book viruses are attributed human characteristics. I make no apology for this. If it helps to bring viruses alive for the reader and reveals them for what they are—ingenious, manipulating, and ultimately harmful parasites—then it has done its job.

In the first chapter I outline the nuts and bolts of viruses and then describe the age-old struggle between virus and host which has helped to shape the destiny of both parties. In subsequent chapters I discuss the various strategies that viruses use to survive and reproduce in their host, many of which cause disease. Each chapter focuses on a particular pattern of infection—be it emerging, acute, chronic, or cancer-causing. In the final chapter I look at past, present, and future treatment options. I use individual viruses as examples throughout, so that some, like HIV, make an appearance in several chapters. I have included 'mad cow' disease in this book because it is caused by an infectious agent of some sort, although we now know that this is not a classical virus.

Many people have assisted in the writing of this book, and it is simply not possible to thank them all here. Indeed most of them will not know that they have helped because over the last few years my kleptomaniac brain has snapped up any interesting facts, figures, phrases, or quotes about viruses, written or spoken, which would serve to illustrate a point. Those who have given expert advice and information on specific topics are acknowledged in the text; here I wish to thank all those who have kindly spared the time to read, correct, and advise on the manuscript. My sincere thanks are particularly extended to Glenda Faulkner and J. Alero Thomas for their constructive comments, Sabine Austin–Brooks for literature research and expert secretarial assistance, and Susan Harrison for her welcome editorial help and advice.

I hope that the book provides interesting and thought-provoking reading.

D.H.C.
Edinburgh, September 2000

This book is dedicated to my mother,
Margaret Crawford (1913–73).

Contents

Introduction

The deadly parasite

It is July 1967. A totally new and unexpected enemy hits US merce-
naries fighting in the rain forest of the remote Motaba River Valley
in Zaire. It kills quickly, efficiently, and horribly. Victims suffer
from blinding headache, soaring fever, and catastrophic bleeding
before their lungs literally dissolve away. Death is a merciful release.
It looks as if the enemy have released a toxic gas, but in fact, a deadly virus has swept through the troops like a tornado, leaving devas-
tation in its wake. A brief exploratory visit by US army chiefs
convinces them they want nothing to do with the situation. They
drop a bomb on the valley to eliminate the evidence; then they suppress the whole episode.

But 30 years later the same virus comes back to haunt them—this
time in the US. Motaba Valley virus arrives in San Francisco by
boat, carried by a lovable and apparently innocuous pet monkey.
Captured in an African rain forest and sold in a market, a small
bribe to US harbour officials is enough to evade the quarantine
period and gain illegal entry into the US. So the deadly virus sets
foot in the country, and after spitting virus-infected saliva at its
owner and scratching a potential buyer with virus-contaminated
claws, the monkey is released into the pine forests of California.

Meanwhile, even with an incubation period as short as 24 hours, the virus manages to travel far and wide while inside its victims. It finds its way to Boston by plane where it infects the partner of the first victim via a kiss. Then, through droplets generated by a cough, it infects an entire cinema audience. From here on we have an epidemic of a virus which spreads like wildfire, causes horrendous suffering, and is one hundred per cent lethal.

Epidemics like this have all the makings of a good thriller—suspense, panic and disaster, and eventually a feel-good ending. This one is the beginning of the film 'Outbreak' released in 1995.[1] Actor–medics Dustin Hoffman and Rene Russo cope with the rapidly developing crisis and, in the nick of time, find a cure which prevents total extermination of the human race. This doomsday scenario is alarming enough, but the really frightening thing is that the film is based on fact—the first real-life outbreak of Ebola virus.

At the beginning of the film a quote from Nobel prize winner Joshua Lederberg flashes across the screen:

> **'The single biggest threat to man's continued dominance on the planet is a virus.'**

Is this true? Are viruses really clever enough to outwit us? How did they earn such an awesome reputation?

This book examines the battle which has raged between man and viruses since he evolved from the ape. How has man coped in the past? Where do new viruses come from? Are we equipped to combat such an attack? Could a new virus wipe out the human race?

I discuss all these questions and then try to answer the vital one—what will the outcome of the battle be? Who will be the final victor—man or microbe?

In the last 25 years, outbreaks of new virus infections have increased dramatically. Lassa, Ebola, human immunodeficiency virus (HIV), and Hanta virus have all appeared as if from nowhere. They have caused devastating disease, producing fear and panic, and have spread rapidly, killing most of those infected. Ebola is a good example. In 1976 a school teacher in the jungle village of Yambuku, Northern Zaire, fell ill. He arrived at the local mission station complaining of headache and fever, so the Belgian nuns at the mission gave him an injection of antimalaria drugs. But he did not have malaria, he had Ebola. Within a few days he was dead, but not before infecting several others with the deadly virus, including his family and the nuns who nursed him. This started an epidemic which affected 318 people, killing 280 of them.

Lassa fever virus in Nigeria and Hanta virus in USA have caused similar recent epidemics and loss of life, but the most terrifying new infection of the twentieth century is HIV. Heralded as an exterminator of the human race, this virus has spread out of Africa to every country in the world in little more than 20 years. It has already killed over 13 million people and shows few signs of stopping. Another 33 million (that is more than the entire population of Canada) are infected and almost certainly condemned to an early death.

Before exploring these viruses in more depth we need to find out exactly what viruses are, how they differ from other microscopic life forms, and why they have gained such a fearful reputation.

Chapter 1

Bugs, germs, and microbes

Microscopic bugs are ubiquitous, the human body is assailed by an invisible army of them. Bacteria, protozoa, and fungi lurk everywhere, apparently just waiting for an opportunity to attack. Indeed, there are more 'bugs' grazing on or in a single human body than there are people in the entire world. But very few cause disease, and some even do us a favour by devouring dead skin cells or helping to break down indigestible molecules in our intestines. In return we provide them with food and shelter—a true symbiotic relationship.

Viruses, in contrast, cannot simply 'graze' on us or any other living thing. They are much more demanding because to survive and reproduce they must penetrate a living cell. Once inside, a virus takes what it needs from the cell and gives nothing in return. So viruses are parasites and although they are often lumped together with other microscopic life forms as 'bugs' or 'germs', they are actually fundamentally different. These differences make viruses unique and sometimes deadly adversaries.

Clever, subversive, subtle, ingenious. These are some of the adjectives commonly applied to viruses and they seem to describe them admirably. But do they paint a true picture of a virus? Certainly viruses exploit every available opportunity and can outwit their hosts sufficiently to get inside and cause disease. They appear to be able to plan an attack and survival strategy, but this is assuming that viruses can think. However, viruses do not have brains and therefore they are not in control of their own destiny. So how can something so small and simple be so smart? How can they take over and wreak havoc in much more complex and sophisticated beings? These are fascinating questions which are a continuing challenge to scientists around the world.

Invisible invaders

Viruses represent life stripped to the bare essentials. They are the smallest and simplest infectious agents identified to date. (Prions (putative infectious proteins) are smaller, but their nature has not yet been definitely established. See Chapter 4.) Indeed they are little more than a piece of genetic material protected by a protein coat (or as Sir Peter Medawar said—'a piece of bad news wrapped up in protein').[1]

The word 'virus', meaning 'a submicroscopic entity',[2] was coined at the beginning of the twentieth century when scientists at last realized that viruses are not just very small bacteria. Most common infections are caused by either viruses or bacteria, but it is not easy to distinguish between the two simply by looking at an infected person. Measles, chickenpox, mumps, and German measles are caused by viruses, whereas diphtheria, whooping cough, typhoid, and TB are bacterial infections. Both can cause pneumonia, meningitis and gastroenteritis. So it is no wonder that viruses and bacteria are more or less synonymous in most people's minds.

The memorable newspaper headline:

'Killer bug ate my face':

appeared in 1994 following an outbreak of a 'flesh-eating bug'.[3] Press reports called the culprit 'a virus', although the disease, necrotizing fasciitis, is actually caused by a bacterium. And again in 1996, when the world's worst outbreak of *E. coli* 0157 (a bacterium) gastroenteritis emanated from John Barr's butcher's shop in Wishaw, Scotland, infecting over 200 people and killing 29 of them, at least one newspaper blamed '*E. coli* virus'.[4] Now, according to the *Independent on Sunday*, even the deadly multidrug-resistant TB bacterium is also a virus.[5]

This kind of misinformation by the press fuels the conviction that viruses and bacteria are similar beasts, but although both are invisible to the naked eye, there the similarity ends. Even their relative size differs enormously; if a virus was the height of a man, an average-sized bacterium would be as high as the Statue of Liberty. In reality bacteria are between 1 and 10 microns in length (one micron is a millionth of a metre), and something like three million of them could sit comfortably side by side on the head of a pin. You can see bacteria with an ordinary light microscope which can magnify 200 times, but viruses, being up to 500 times smaller, can only be seen with an electron microscope which can magnify 100 000 times.

Bacteria are direct descendants of earliest life on earth, and represent the blue print for cells which make up all other animals and plants. They are the smallest microbes which can survive without help from any other living thing, most of them being just single cells which live in the natural environment. Life on this planet could not exist without bacteria which act as miniature recycling plants for atoms and molecules. Every day they are busy breaking down dead plants and animals into their component parts ready for reuse. The gases released make up our atmosphere, and simple molecules are rebuilt into new plants and animals. Only rarely do bacteria invade other living things and cause disease.

Bacteria grow and divide in a similar way to cells in more complex beings, that is by simply splitting in two. Provided that they have all the nutrients they require, bacteria can divide once every 20 minutes or so. Hence, within 24 hours a single bacterium can produce a colony of over 4 000 000 000 000 000 000 000 (4×10^{21}) identical bacteria.

Like all living things, bacteria contain genetic material—DNA— which ensures that their characteristics are inherited by their offspring. Their DNA carries the information to make approximately 4000 proteins which is enough for the bacterium to live and reproduce on its own. Bacteria also contain all the necessary cellular machinery to read DNA messages, download the information into RNA, and produce proteins from it. All this activity is powered by complex metabolic processes which require an input of oxygen and release carbon dioxide.

Unlike bacteria, viruses can do absolutely nothing on their own. They are not cells but particles and they have no source of energy or any of the cellular machinery needed to make proteins. Each particle simply consists of genetic material surrounded by a protective protein shell called a 'capsid'. These particles brave the outside world to carry the genetic material of a virus from cell to cell and at the same time spread the infection. Viruses only have between 3 and 400 genes (humans have about 30 000) and these tiny pieces of genetic material carry the code for their own reproduction. But in order to reproduce they have to penetrate a living cell and take control.

Nowadays the term 'virus' is used to describe non-biological agents like computer viruses. These modern 'viruses' are invisible parasites which infect, reproduce and cause disease, albeit in computer programs rather than living cells. They are transmitted by floppy disks or e-mail messages, and once inside a computer they spread, disrupt programs, and cause chaos. Treatment packages are of limited use because viruses 'mutate' so frequently. In all these characteristics computer viruses closely parallel their biological counterparts. In the analogy, biological viruses are the floppy disks which only function when they have access to computer hardware which is the cellular machinery. As long as a virus can get inside a cell, the cell will read the virus' genetic code saying 'reproduce me' and get on with the job.

So viruses invade living things, commandeer their cells, and turn them into factories for virus production. In a day or two, thousands of new viruses emerge. Viruses are not intent on causing disease, but infection usually weakens or destroys cells, so if enough cells are infected it usually has an effect. Whole organs can be wiped out and if they are vital and irreplaceable then infection will be fatal. Rabies virus, for example, destroys brain cells and Ebola kills cells which line blood vessels, causing catastrophic bleeding.

Some viruses allow infected cells to survive, albeit in a weakened state, and continue to produce new viruses. When this happens the infection can have bizarre effects on the cell and animal or plant as a whole. A famous example of this caused 'tulip mania' in Holland in the seventeenth century, when beautiful variegated tulip flowers were first cultivated. Tulips were brought to Europe from Turkey in the mid-sixteenth century and Holland soon became the centre of the tulip trade. From a plain red beginning, plant breeders developed variegated flowers with white strips called 'colour breaks'. These exotic blooms were white at the base with a delicate filigree of white on red patterning the petals (Fig. 1.1). 'Broken' tulips were

Fig. 1.1 The variegated 'broken' tulip first cultivated in the seventeenth century.

rare and much prized as a status symbol. But the flowers were as fickle as they were beautiful. As Anna Pavord relates in her book *The Tulip*, 'The flower had a unique trick which added dangerously to its other attractions. It could change colour, seemingly at will'.[6]

Although once 'broken' a bulb remained so, only one or two in a whole field might show the effect. None of the fascinated plant breeders could work out why the colour suddenly changed. It certainly did not follow the rules of genetics learnt from the study of other plants. Those with colour breaks were less vigorous than their plain-coloured counterparts, and all this conspired to make the bulbs expensive. But between 1634 and 1637 things got completely out of control and tulip mania took hold. A single prize 'Admiral van Enkhuijsen' bulb sold for 5400 guilders (around £400 000 in today's money)—the cost of a town house in smart Amsterdam and the equivalent of 15 years' wages for a labourer. At the height of the mania it was cheaper to commission a famous artist like Jan van Huysum to paint you a picture of a 'broken' tulip (for up to 5000 guilders) than to buy a single bulb.

The early growers, who had plenty of theories on how to encourage colour breaks, could not have known that it was all caused by a virus. The tiny parasite suppresses colour formation and at the same time weakens the plant. Aphids which inhabit fruit trees spread the virus, and since fruit and tulips were often grown together in Holland it is not surprising that the infection regularly made its appearance, but in an inexplicably random way.

From miasma to virus

Louis Pasteur and Robert Koch were the first to show that infectious diseases were caused by microbes. Working in the middle of the 1800s—Pasteur in Paris and Koch in Berlin—these rivals between them identified bacteria in diseased material, developed methods for growing them in culture, and showed that they could transmit disease to susceptible animals. Anthrax bacillus, which primarily causes fatal disease in sheep but also infects humans, was the first to be isolated. This was followed by many more discoveries including the causes of killer diseases like TB and cholera. These dramatic advances led to the modern 'germ theory' of infectious disease.

From the time of Hippocrates until the beginning of the nine-teenth century, diseases were generally thought to be caused by two forms of poison: 'virus' and 'miasma'. 'Virus' referred to visible poisons like snake venom, the saliva of rabid dogs, and poisonous secretions of plants. 'Miasma' was an invisible gas which emanated from swamps, stagnant water, unburied human bodies, and animal carcasses and caused infectious diseases and plagues. It required an enormous leap of faith for a population not trained in scientific thinking to relinquish these beliefs and accept that diseases were actually caused by tiny living things which colonized their bodies from within rather than by noxious substances coming from the outside. Not surprisingly, for some time after the revolutionary discovery of germ theory, it was still generally believed that bacteria were the result rather than the cause of a disease, and it was not until the middle of the nineteenth century that the theory was universally accepted.

By the beginning of the twentieth century several hundred bacteria had been identified, but there were still many severe infections for which no cause could be found. Several of these were common diseases such as measles, smallpox, rabies, yellow fever, as well as foot and mouth disease of cattle, and tobacco mosaic disease of plants. Because infection had devastating effects on human lives, all these diseases were studied in great detail, but no bacteria could be isolated.

In 1876, Adolf Mayer, a German scientist and Director of the Agricultural Research Station at Wageningen, Holland, was probably the first to transmit a virus disease successfully. He worked on tobacco mosaic disease, which he so named because of the dark and light spots it causes on the leaves of tobacco plants (Fig. 1.2). The disease was of great economic importance to the Dutch as it could

Fig. 1.2 Leaf of nicotiana (tobacco plant) showing mottled effect of tobacco mosaic virus infection. (Reproduced from *Introduction to modern virology* by S.B. Primrose and N.J. Dimmock, 1980, courtesy of Blackwell Scientific Publications.)

devastate their lucrative tobacco crops. Mayer mashed up leaves from infected plants and spread the disease by rubbing the extract onto leaves of healthy plants. He thought that the active ingredient must be an enzyme or toxin. Then a Dutch soil microbiologist working with Mayer, Martinus Beijerinck, further clarified the mystery by passing diluted juice from infected leaves through porcelain filters with pores small enough to retain all known bacteria. The filtered juice not only remained infectious but also regained its original strength after infection of a second plant. This proved that the agent could reproduce and was therefore a living micro-organism of some sort, rather than an enzyme or toxin.

There followed a period of intense and heated debate over the nature of these so-called 'invisible microbes' which lasted until 1903 when Pierre Roux, a successor of Pasteur as Director of the

Pasteur Institute in Paris, spelt out their properties in scientific terms. He defined three measurable characteristics:

1. Filterable—because of their small size they could pass through filters which retained bacteria.

2. Invisible—they could not be seen under a light microscope.

3. Non-culturable—they would not grow on bacterial culture plates.

The term 'filterable virus' was coined to describe these microbes, but even so for the next three decades viruses were regarded by most scientists merely as very small bacteria.

Between 1900 and the 1930s the physical nature of viruses became clearer, but their methods of infection and reproduction remained a mystery. Refinement of filter pores meant that their size could be determined accurately, and new cell culture techniques allowed many viruses to be grown in cultured cells, purified, and then shown to cause disease in living animals. Eggs containing chick embryos turned out to be a convenient place for growing certain viruses because inoculation on to membranes surrounding the developing chick gave a visible effect (a plaque) after three to four days. Microscopic examination of these plaques led to the realization that viruses multiplied *inside* living cells, and generally killed the infected cell.

Invention of the electron microscope in 1938 finally gave clear images of the submicroscopic world of viruses. At last the incredible structures and symmetries of these tiny microbes could be studied for the first time (Fig. 1.3).

But opinion was still divided on the nature of viruses. Crystallization of tobacco mosaic virus in 1935 seemed to prove that it was pure protein which was in some way self-renewing.

However, after the discovery that chromosomes carry genes which are composed of DNA, and the final cracking of the double helix structure of DNA by James Watson and Francis Crick in Cambridge in 1953, it eventually became clear that viruses contain genetic material as well as protein. This finally placed viruses in a category of their own, and the essential differences between viruses and bacteria were fully appreciated.

Alive or dead?

The Oxford Dictionary defines life as 'the condition which distinguishes active animals and plants from inorganic matter, including the capacity for growth, functional activity and continual change preceding death'.[2] This is a rather unhelpful definition as far as viruses are concerned because they only become active after they penetrate a living cell. Although they possess some characteristics of living things, they lack many features essential for life. For instance, they can reproduce, albeit only with a great deal of help. On the other hand, they lack all the metabolic processes needed to generate energy, and the molecular machinery required to make proteins.

It is pertinent at this point to ask where viruses came from. Did they, for example, evolve from some other living thing and if so, from what? Unfortunately, viruses have left no fossil records by which to track their origins. But although they were only discovered about a hundred years ago, we know that they have been around far longer than the human race. Practically all living things, whether plant, animal, or microbe, have their own particular viruses which have evolved with them over millions of years. Scientists interested in the origin of viruses look for clues among today's viruses by

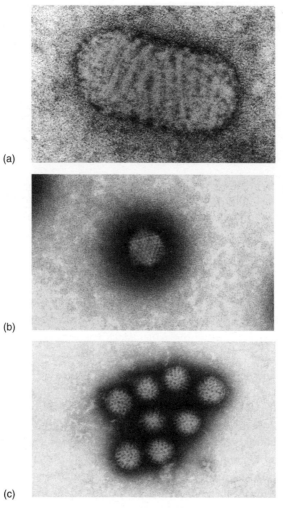

(a)

(b)

(c)

Fig. 1.3 (a)–(c) Electronmicrographs of: (a) pox virus (orf) (×180 000); (b) adenovirus (×180 000); (c) rotavirus (×130 000).

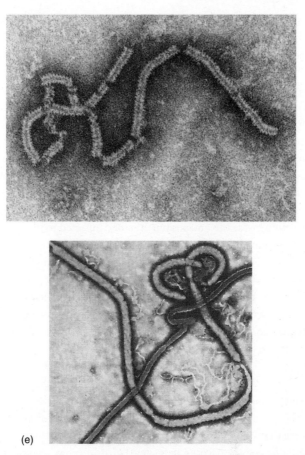

(d)

(e)

Fig. 1.3 (d)–(e) Electronmicrographs of: (d) mumps virus (nucleo-protein helix, ×180 000) (All courtesy of Dr Hazel Appleton, Central Public Health Laboratory); and (e) Ebola virus (×24 000) (C.J. Peters *et al.* (1996) Filoviridae: Marbury and Ebola viruses. In *Virology* (ed. Fields) Third edn. Lippincott-Raven Publishers, with permission)

comparing their genetic make-up with that of other microbes. Just as man's evolutionary tree can be traced back through look-alike apes and monkeys to the simplest mammals, there is a series of microbes ranging from the most sophisticated free-living bacteria down to the simplest molecules which can reproduce.

These days most people believe that viruses are rogue pieces of genetic material which have somehow broken free from chromosomes and found a way to reproduce independently. 'Jumping genes' may represent the beginning of this process. They can free themselves from the DNA chain of a chromosome and rejoin at another site, but they cannot get out of a cell. One step further on from this are plasmids. These fragments of DNA are found in bacteria living inside the nucleus but are not connected to the bacteria's chromosomes. They carry extra genes which are sometimes harmful to the host. For instance, plasmids can provide the genes for making bacteria either resistant to antibiotics or produce harmful toxins. Like viruses, plasmids rely entirely on the cell for their reproduction, but they are not classified as viruses because they do not make particles. Consequently they are trapped inside a cell and their ability to spread is severely limited. Mostly they just pass from parent to daughter cell when the cell divides, but occasionally they are transferred directly from one organism to another during conjugation when two bacteria join briefly to swap genetic material.

Plasmids may be the forerunner of viruses but some people believe that viruses are derived from bacteria which regressed from free-living cells to become parasites. For this to have happened, at some point in the distant past they must have found it easier to steal essential nutrients than to make them themselves, eventually losing their production ability altogether and so becoming obligate parasites. However, this scenario seems unlikely because although degenerate parasitic bacteria do exist, they obviously resemble bacteria and are quite distinct from viruses.

So viruses could either be descendants of living bacteria that have taken parasitism to its extreme, or simply pieces of genetic material with a built-in code for their own reproduction. Thus the answer to the question of whether viruses are alive or not remains a matter of opinion and personal preference. Most medical specialists in infectious disease think of an infectious agent with inherited characteristics as living, whereas many molecular scientists tend to regard viruses as just another molecule of genetic material which can be manipulated in a test-tube. In the long run does it really matter? Whichever way you look at it, there is no doubt that viruses are unique and utterly unlike cells which are the building blocks of all other organisms.

The troops gather

Every time a virus infects a host a battle ensues. Flu virus, for example, has just three to four days to establish an infection before its host either dies (fortunately a rare event) or controls the infection and eliminates the virus. In that time the virus has to infect as many cells as it can and reproduce as quickly as possible; its offspring must exit before they are destroyed by the host's immune system. Whatever the outcome of the encounter, sooner or later a virus must move on. Many viruses can only survive for one or two days outside a cell before drying out and becoming inactive, so after release from one host they have to find and invade another quickly to maintain an unbroken chain of infection and guarantee their long-term survival.

Viruses cannot take an active part in their spread because their particles are completely inert. They have to take their chances in the outside world, drifting on air currents, floating in liquids, and lurking in food. Like the seeds of plants, many millions are produced in the hope that at least one will find and take root in a new host. Viruses exploit every possible avenue to colonize a new host and their success is self-evident.

Viruses generally gain access by infecting cells on body surfaces—either skin or internal surfaces of the intestinal, respiratory, and genitourinary tracts. Here they produce masses of new viruses which are ideally placed either to spread to internal organs or to hitch a ride in secretions or excretions back to the outside world.

As children most of us had harmless but unsightly warts on our hands or painful verrucae on the soles of our feet. Papilloma virus is the culprit here, penetrating our skin through a small cut or abrasion, and once there infecting developing skin cells. To aid their own reproduction, these viruses stimulate the infected cells to grow more rapidly than surrounding normal skin cells, the extra cells pushing up into the familiar tiny cauliflower-shaped lump of a wart which in turn sheds more viruses. A mere handshake may then be enough to complete the cycle of infection with its spread to other people.

Herpes simplex virus also infects skin cells making them swell up and burst, releasing their cargo of new viruses, and causing a painful rash of tiny blisters. On the face, usually somewhere near the lips, this causes the common cold sore, and since the blisters are full of viruses, infection is easily spread by a kiss. In the genital area, herpes simplex virus causes genital herpes—the commonest type of genital ulcer. This is also highly infectious and easily spread by unprotected sexual intercourse. But spread by sexual contact is not restricted to viruses which grow in the genital tract; viruses in the blood, like hepatitis B and HIV, also take advantage of this route. Genital secretions normally contain a few white blood cells which may carry viruses, but just a minute amount of bleeding from injury or from genital ulcers is enough to increase the chance of sexual transmission of blood-borne viruses quite dramatically.

'Delhi belly' and 'raspberry desert'

Most of us have been mildly inconvenienced on our travels by 'tummy bugs' (known graphically as 'Delhi belly', the 'Aztec two-step', and so on), but these infections can be serious. Rotaviruses alone kill some 800 000 people every year. They thrive in the overcrowded and unhygienic conditions common in developing countries where they are usually spread from person to person by contaminated food and drinking water (the faecal–oral route). They are adept at surviving their passage through the acid environment of the stomach, and then grow in the small intestine. Here they wipe out the cells of the gut wall causing acute diarrhoea and vomiting. Massive loss of fluids results in dehydration which rapidly kills the weak and undernourished. It also flushes out the virus so that there are something like 1 000 000 000 (10^9) rotaviruses in every gram of faeces passed by an infected person. And these can survive for several weeks in sewage and water while waiting for an opportunity to infect a new host.

But food poisoning is not confined to developing countries. When 120 eminent doctors gathered for their 1977 annual dinner at the Apothecaries Hall in London, they little thought what the consequences might be. Two to three weeks later 50 guests, as well as the cook and a waiter, came down with jaundice. The medical detective on the case, Norman Noah, identified the culprit as hepatitis A virus, and the carrier was the dessert they had all eaten—raspberry parfait.[7] The raspberries, he discovered, had been picked in Dundee, Scotland, two years before and frozen at a local factory. Noah suspected a single batch of about 20 contaminated punnets, and decided that the source was either an infected factory worker who had handled the raspberries briefly while adjusting their weights or an unidentified picker in the fields.

This outbreak was bad enough, but during the investigation some startling practices among raspberry pickers came to light. Raspberries picked for freezing must be dry and perfect, so they are placed in small punnets. But raspberries picked for jam do not have to be in such prime condition and they are collected in large barrels. Pickers out in the fields developed the habit of urinating into these barrels because it increased the weight and thereby their pay. An unpleasant habit certainly, but fortunately not dangerous since any contaminating viruses would be killed by boiling the fruit during jam making.

The old saying 'coughs and sneezes spread diseases' is spot on; a sneeze generates a fine spray of fluid secretions, like an aerosol from a pressurized can. Since viruses which cause colds and flu grow in the nose and sinuses, where they stimulate the sneeze reflex, it is not difficult to imagine how they are spread. Millions of virus-laden droplets from an infected sneezer float in the atmosphere ready to be inhaled by cohabitants of a crowded train, classroom, or nursery. This is the most common and efficient method of virus spread in temperate climates and is the route used by the most successful viruses like measles, mumps, rubella, and chickenpox. These viruses infected nearly every child before vaccination put a stop to them.

Blood suckers

The best-known microbe transmitted by a mosquito is probably the malaria parasite, *Plasmodium,* but viruses also exploit blood-sucking insects to help them out. These viruses primarily infect animals and are carried from one host to another by ticks, sand flies, or mosquitoes. Infected animals act as a natural reservoir of infection, and anyone who happens to encroach on the animal–insect–animal cycle and be bitten by a virus-carrying insect may get infected. Yellow fever and dengue fever viruses are the commonest insect-borne viruses which cause problems for humans. They both naturally infect monkeys, are both transmitted by the female *Aedes aegypti* mosquito, and both cause deadly haemorrhagic fever. Because their life cycles are intimately linked to that of the mosquito, wherever she thrives so do the viruses. But this link, which was first identified for yellow fever, took some time to unravel.

Yellow fever was first reported in Havana, Cuba, in 1648 and it soon took hold in South and North America. The virus was probably brought to America from its roots in Africa aboard slave ships bound for Barbados in around 1647. Mosquitoes survived on board by breeding in the ship's water barrels and continued to spread the virus among passengers and crew during the long voyage. Once in port, mosquitoes moved inland and established themselves as natives of the local rain forests.

In 1898, during the Spanish–American war, an epidemic of yellow fever struck the US forces occupying Cuba and 231 soldiers died as a result. This catastrophe finally persuaded the American government to take action against this much-feared disease which had by then plagued the Americas for over two centuries. Victims developed a high fever, severe head and joint aches, bleeding, nausea, and bloody vomiting. Many died at this stage from internal bleeding but others went on to develop the jaundice which gives the disease its name, often dying of liver, heart, or kidney failure. Epidemics used to occur regularly in riverside towns with a warm, wet climate like Memphis and New Orleans and also in ports like New York and Philadelphia. A devastating outbreak in Philadelphia in 1793 killed over 4000 people—a tenth of the population—in just four months.

The way yellow fever spread was a mystery. It did not seem to pass directly from person to person, or contaminate food or drinking water, but it was often carried from one community to another by people fleeing a disease-ridden town. Its appearance in ports often coincided with visiting ships, and yet putting ships in quarantine did not prevent disease spreading to the town. No bacteria could be isolated from infected people, and all this confusion led some to believe that yellow fever was not an infectious disease at all.

In 1900, the army set up the US Yellow Fever Commission and four American doctors led by Walter Reed, Professor of Bacteriology at the US Army Medical School, headed for Havana to investigate this killer disease. They carried out experiments on volunteers to find out if mosquitoes spread the disease, and three of the group offered to act as human guinea pigs. They eventually convinced themselves that the virus was spread by mosquitoes, but not before two of the party, James Carroll (originally Reed's laboratory assistant) and Jesse Lazear (head of the clinical laboratories at Johns Hopkins Medical School) contracted the disease. Carroll became seriously ill but eventually recovered, while Lazear tragically died.

The idea that mosquitoes could spread diseases was entirely new at the time and at first it was ridiculed. However, after further carefully controlled experiments with more human volunteers enough people were convinced to start organizing preventive measures. In 1901, a team from the US army under Major William Gorgas set about ridding Havana of mosquitoes by removing their breeding grounds. They either drained all standing water or covered its surface with oil, and this successfully controlled yellow fever in the city. Finally, a vaccine was developed in 1937 which has prevented major outbreaks ever since.

Monkeys of the African and South American rain forests act as reservoirs of yellow fever virus by carrying it in their blood. However, they remain healthy and fully active, probably because over the years their bodies have got used to the infection. Once she has taken her bloody meal from a yellow fever virus-carrying monkey, a female mosquito injects the virus into her next victim, whether it be monkey or human. But the virus is not just a passive traveller in the mosquito; it multiplies in her intestine and travels to the salivary glands where it also multiplies whilst waiting to be injected with saliva into the next victim. So the virus amplifies several thousand times without causing any illness to the mosquito which might jeopardize its chances of successful transmission. Once mosquitoes have transmitted yellow fever from monkey to human they can then carry it directly from one human to another, and will do so readily in crowded conditions (Fig. 1.4).

Since mosquito-borne yellow fever virus successfully crossed the Atlantic, it is surprising that it has not yet managed to cross the Indian Ocean and establish itself in Asia. However, the risk of this happening is higher now than ever before. The disease, which for years was confined to West Africa, has recently spread to Kenya in the East. Constant traffic in the form of cargo ships plying between Kenya and India provides ideal transport across to another continent where the virus would find a wholly unprotected population to infect.

In contrast to yellow fever, dengue fever virus is rife in both Africa and Asia, and there is no vaccine to control it. Although the virus often only causes flu-like symptoms, dengue can produce a

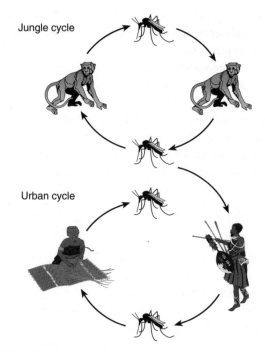

Fig. 1.4 Yellow fever transmission cycle. In the jungle the virus is carried from monkey to monkey by a mosquito. If a virus-carrying mosquito bites and infects man, then there is the potential for human to human spread.

potentially fatal haemorrhagic fever. A recent outbreak in Delhi led to 5500 hospital admissions and 320 deaths in just two months. Epidemics like this have increased dramatically of late in both Africa and Asia where rapid, unplanned urbanization provides the overcrowded conditions in which the virus thrives. In these cities, virus-carrying mosquitoes breed anywhere—in ponds, puddles, drains, drinking water reservoirs, and air conditioners. Inside the rim of old rubber tyres is a particularly favourite breeding ground, and there are plenty of those in any large city.

During each monsoon season mosquitoes undergo a population explosion, and the virus which they spread reaps the benefit. Yellow fever and dengue viruses are at the mercy of their vector, and at the moment their outlook is rosy. If global warming continues as predicted then the traditional territory of tropical and subtropical mosquitoes will extend north and south, and the viruses they carry will spread with them to new and virgin territories.

Man-made spread

Viruses quickly exploit modern medical practices. From one virus carrier they can find their way into new hosts in transfused blood, blood products, and organs for transplant, or they contaminate needles, surgeons' knives, and dentists' drills. The main culprits are blood-borne viruses like hepatitis B and C and HIV. These all establish chronic infections in apparently healthy people who are often unaware of their presence. Of the three, hepatitis B is the most infectious; each millilitre of the blood from a carrier may contain several million viruses, so infection can be transmitted by a microscopic amount.

These days blood is screened before it is deemed safe to use, so that the chance of catching anything from a transfusion is extremely small; on average around 1 in every 34 000 units of blood is contaminated in the UK. But still, people who have regular infusions, like haemophiliacs requiring blood-clotting factor VIII, are particularly at risk. It was this that caused an HIV epidemic among haemophiliacs in the US and Europe in the 1980s (which is discussed further in Chapters 2 and 4). And the early spread of HIV in the 1970s in Africa was probably facilitated by the reuse of needles for smallpox vaccinations that were not sterile.

Hepatitis C virus is most often spread by unsterile needles, and so is common among intravenous drug users. Like hepatitis B, chronic infection with hepatitis C causes liver damage which may lead to cirrhosis or cancer (see Chapter 5). This is a real public health problem, particularly in developing countries where in some areas up to 10 per cent of the population are symptomless carriers. A recent outbreak of hepatitis C virus in Vallencia, Spain, which infected 217 people, was eventually traced to an unusual source.[8] All those infected had undergone a surgical operation in one of two local hospitals, the state-run La Fe or the private clinic, Casa de Salud, in the previous two years. The link was a doctor, Juan Maeso, who gave the anaesthetic in every case. He turned out not only to be infected with hepatitis C but also a long-standing morphine addict. Part of his job was to give patients an injection of painkillers immediately after their operations, and while filling the syringe with morphine he conveniently helped himself to a shot before administering the rest to the patient using the same needle, now contaminated with his hepatitis C-infected blood.

Mother to child

The Australian ophthalmologist, Sir Norman Gregg, working at The Royal Alexandra Hospital in Sydney, was the first to discover that viruses could spread directly from mother to unborn child. In 1941 he noticed an unusually high number of babies born with cataracts and heart abnormalities and he related this to German measles (rubella) infection in the mothers who had been pregnant during an epidemic the previous year.[9] Viruses in the mother's blood may cross the placenta, grow in the baby, and damage developing organs. The earlier in pregnancy that German measles occurs, the higher the risk to the baby, and the more severe the damage. Infection during the first month of pregnancy, when the baby's organs are developing, affects almost all babies, whereas after the fourth month, when the baby is fully formed, no abnormalities occur. Of course only those mothers who had escaped German measles as a child were at risk of catching it during pregnancy, and nowadays the risk can be removed by rubella vaccination.

As well as spreading directly from mother to child through the placenta before birth, viruses like hepatitis B in the blood and herpes simplex in the birth canal can infect the baby during birth, while others, like cytomegalovirus and HIV, cross from mother to child in breast milk.

The battle commences

We have seen how viruses travel between their hosts, but finding a new susceptible host is only the beginning of the story. They then have to penetrate host cells before they can come to life, and here again viruses are remarkably resourceful. Normal cells are bathed in a sea of chemicals, like hormones and growth factors, all of which are looking for a way to get inside. But entry is restricted by a series of receptor molecules on the cell surface which act as locks each of which can only be opened by the particular chemical whose molecular key fits it precisely. This restricted entry ensures that each type of cell behaves appropriately—that only nerve cells respond to nerve cell growth factor, T cells to T cell growth factor, and so on. But this mechanism also provides viruses with a route of access. By

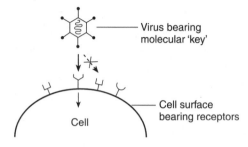

Virus bearing molecular 'key'

Cell surface bearing receptors

Cell

Fig. 1.5 Virus infection of a cell. A virus uses its 'key' to enter a cell via a particular receptor molecule on the cell surface.

carrying a molecular key on their surface, they can disguise themselves as normal body constituents, latch on to, and enter, any cell which bears the complementary lock (Fig. 1.5).

So viruses infect only those cells which display the particular molecular lock which their key fits, and this restriction dictates the type of disease a particular virus will cause. Since there are several hundred molecules to choose from, viruses cause a great variety of diseases. A well-known example is HIV which carries the key for a cellular molecule called CD4. It therefore infects and destroys CD4 positive cells. As we will see in Chapter 4, these cells are central to the functioning of the immune system and when they are destroyed by the virus faster than they can be replaced, the body's immune defences crash and acquired immunodeficiency syndrome (AIDS) is the result.

So far we have looked at infection from the viruses' point of view—how they invade a host, penetrate cells, and commandeer them for their own ends. The ingenuity viruses display is astonishing, but they are not fighting a one-sided battle. Even the simplest organisms have ways of dealing with viruses, but the sophistication and subtlety of the human immune system is unrivalled.

As soon as we are born, and every day of our lives, our bodies are like castles under siege, surrounded by swarming enemy troops all trying to breach the walls, enter, and plunder the contents. Each enemy carries a weapon to assist it and each tries a different port of entry. But just like fortified castles, we are built to withstand the attack.

The first strategy is to prevent access. For this we are covered with a thick outer skin which, when intact, is impenetrable. It comprises many layers of brick-like cells all joined together by interlocking processes and covered with a layer of flattened dead cells. Viruses cannot infect dead cells because they are metabolically inactive, so they must enter by injection or injury, or through one of the natural orifices.

Although our outer skin is continuous with the gastrointestinal, respiratory, and genitourinary tracts, these inner linings have no protective covering of dead cells and are often just a single cell thick. Here viruses (and other harmful microbes) can get a foothold, but we can usually forestall this line of attack. Secretions like tears and mucus, which contain antiseptic substances, trap and evict invaders, while in the vagina and stomach, acid secretions destroy all but the hardiest of attackers. Cells lining the upper respiratory tract carry small hairs which beat in unison, acting like an escalator to carry foreign particles up and out. Invaders which successfully bypass all these traps are gobbled up by specialized cells called macrophages (meaning 'large appetite') which tour the tissues of the body engulfing and destroying foreign particles. Once they have devoured an invader, these cells send off a variety of chemical signals which increase blood flow to the area and send the rest of the forces, in the form of B and T lymphocytes, scurrying to the scene.

Like red blood cells carrying oxygen to tissues, B and T lymphocytes patrol all the far-flung corners of the body by travelling along arteries and veins. Unlike other organs such as the liver, brain, or kidney, these immune system cells are not gathered in one place. Occasionally they stop off to communicate with each other in lymph glands but otherwise they are spread all over the body. Lymph glands are strategically placed to guard danger zones where microbes are most likely to attack. Tonsils and adenoids protect the entrances to the lungs and the intestine, while glands in the groins and armpits provide a stockpile of lymphocytes to guard the limbs.

B and T lymphocytes play vital roles in the body's defences. Their importance is amply demonstrated by rare genetic accidents in which one or other is absent or non-functional. Babies born without B lymphocytes cannot make antibodies. These children have no particular trouble combating viruses but have major problems with bacterial infections unless they are given regular antibody infusions. In complete contrast to this, babies born with no T lymphocytes have no trouble with bacteria, but suffer devastating virus infections; they quickly die without a bone marrow transplant. This natural accident tells us that T lymphocytes are vital for protection against viruses. Antibodies can control the spread of viruses in the body, but it is those T cells known as 'killer cells' which, in particular, search out and destroy virus-infected cells before they have time to unload their cargo of new viruses.

It is interesting to compare microbial infection with acute poisoning. An overdose of aspirin, for example, causes a metabolic upset which continues until the drug is neutralized by the liver. Overdoses of the same drug taken a week later, and again the week after that, will have the same effects; however many times you overdose on aspirin your body will never get used to it. But this is not the case with infections because the immune system has a memory. After you have had an infection with, say, measles virus, you will be immune to it for the rest of your life. As we will see in Chapter 3, even infections like flu and the common cold, which seem to recur with monotonous regularity, are caused either by different viruses which give similar symptoms, or by the same virus which has mutated slightly to escape the effects of immunological memory.

To cope with the huge number of different microbes, the human body produces an estimated 50 000 000 000 (5×10^{10}) B and T lymphocytes every day. Each lymphocyte can only recognize one

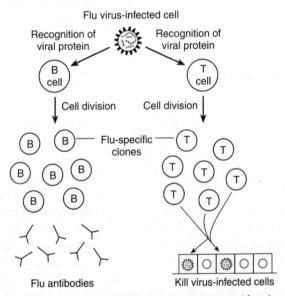

Fig. 1.6 Clonal expansion of B and T lymphocytes. Contact with a virus causes specific B and T cells to multiply. B cells produce antibodies against the virus whereas T cells can kill virus-infected cells.

particular foreign protein, and most are destined to lead a short and useless life, dying without ever meeting the enemy. But if one happens to meet its particular protein, say a flu protein on the surface of a cell infected by flu virus, then that particular lymphocyte is galvanized into action. It divides rapidly to form a clone of identical daughter cells—B lymphocyte clones produce antibodies which bind to and neutralize the flu virus, while T clones kill any flu virus-infected cells (Fig. 1.6). The first time you are infected by a virus it takes five to ten days for these clones to get going, so the virus has time to multiply sufficiently to cause symptoms. But clones of B and T cells, once established, survive a lifetime and are ready to react quickly and prevent disease the next time the same virus comes along.

An evolving killer

Clearly the human body is a battleground, with viruses attacking and the immune system defending on a daily basis; thankfully in most cases the immune system wins out. But now we need to assess how this daily battle influences each of the combatants in the longer term. This means looking back over millions of years to see how viruses and their hosts have co-evolved.

The evolution of all microbes and the diseases they cause has paralleled, and been shaped by, development of efficient ways of preventing and fighting them in their hosts. So as the army of microbes has evolved more and more ingenious methods of attack, the human body has retaliated with further refinements to its defences, so raising the struggle for survival to an unbelievably sophisticated level.

The ancient family of herpes viruses are incredibly widespread in nature—even primitive invertebrates like oysters have their own strains. These and other viruses have diverged and evolved with their hosts over a similar time frame. So when mammals diverged from their reptile ancestors some 220 million years ago, herpes viruses were already in residence. Around 80 million years ago mammals evolved into today's modern species, and contemporary herpes viruses followed their lead. Hence herpes viruses of man are most directly related to herpes viruses which infect man's closest relatives—primates of Asia and Africa—and they have diverged from each other to the same extent as individual primates have.

On an evolutionary time-scale, genes of both host and virus have sustained many mutations (that is, random mistakes which occur approximately once in every hundred thousand to a million times the genetic code is copied). This can alter the nature of a protein, and although most mutations are harmful, occasionally the change is advantageous and the mutated organism becomes more success-ful than its rivals. This is the basis of natural selection which, over millennia, has so exquisitely tuned viruses to their respective hosts that, as we noted earlier, it often looks as if their survival strategies involve detailed planning rather than random trial and error.

In the short term, viruses generally have the advantage over their hosts because they have a much shorter generation time. Viruses produce thousands of offspring every one to two days whereas (Western) man only has an average 2.4 descendants in 20–30 years. Assuming that their mutation rates are similar, it is obvious that more complex organisms will take longer to adapt to a virus than vice versa.

A clear-cut illustration of adaption comes from the man-made epidemic of myxomatosis in rabbits. This virus, which naturally infects Brazilian rabbits and causes them very little harm, was intro-duced into Australian (European) rabbits in 1950 in a deliberate attempt to control their ever-increasing numbers. It had the desired devastating effect; in the first year it killed 99.8 per cent of infected rabbits, but the effect soon wore off. Rabbit numbers reached rock bottom three years after release of the virus, and seven years later only 25 per cent of infected rabbits were dying. Now, some 50 years later, rabbits are back to full strength in Australia (and another virus is being recruited which is discussed in Chapter 2). Since the generation time of a rabbit is only 6–10 months, compared to man's 20–30 years, in a similar situation it would take humans 120–150 years to adapt to a new killer virus.

Evolution and disease

Why were Australian rabbits not wiped out by myxomatosis? Because the rabbits adapted to the virus, and the virus became less deadly. It is often assumed that this type of co-evolution of virus and host invariably leads to milder disease over time since it is usually to the advantage of both. But this is not always the case. It is certainly an advantage to the host, where survival and reproduction are the main driving forces for evolutionary change. Because the most susceptible hosts are the least likely to reproduce successfully, populations which are continually infected with a virus build up a natural resistance over generations which reduces the severity of the infection. Equally, it is disadvantageous for a virus to kill its victims too rapidly as it will soon run out of new ones to infect and its long-term success will be threatened.

Some virus infections cause no disease at all and others kill rapidly, but between these two extremes there is a spectrum of possibilities. The best strategy for each virus depends on its method of spread, incubation period, site of infection, time taken to reproduce, and symptoms it causes. So, for example, a virus which is spread by aerosol, like the common cold virus, needs to be inhaled quickly by a new host before its transporting droplet dries out. Therefore it spreads best in crowded conditions. By only causing mild symptoms of rapid onset and short duration those infected do not retire to bed but continue to meet, and inadvertently infect, as many people as possible during the short time available between infection and the immune response putting a stop to production of new viruses.

In contrast, as we have already seen, yellow fever virus employs mosquitoes for carrying it from one victim to another, and relying on an insect vector is a precarious existence for a virus. So, on the basis that the longer the insect feeds, the more likely it is to pick up a good dose of virus, the best strategy in this instance is to confine the victim to bed, too ill to move around. This gives the mosquito time to feed undisturbed. Here, isolation is not a problem because the mosquito keeps the virus alive for several days while transporting it anywhere within its flight range of up to 2 kilometres.

Viruses spread by sexual contact exploit the basic requirement for survival of a species. In doing so you might imagine that they cannot fail to survive themselves. However, this is not an efficient method of spread in a society which is mainly monogamous. One infected individual will spread the virus to their partner, and there the trail will end. So there is pressure on this type of virus to increase its potential for spread in order to maximize its chances of survival. Most viruses which are spread by sexual contact, like herpes simplex virus, manage this very successfully by evading the immune response and remaining in the body for long periods as a silent infection. Periodic bursts of new virus production over the lifetime of the host then allows this virus the greatest potential for spread to new sexual partners.

Man's most recently acquired sexually transmitted virus, HIV, maximizes its chances of success in a similar way. It establishes a silent infection during which the person remains well and sexually active, but also infectious, for many years. Although it seems likely that HIV will become less virulent with time, the virus is evolving rapidly and virulence may still get the upper hand. If, for example, the average number of sexual partners per person increases, then a more aggressive strain of HIV, with a higher rate of virus production (and consequently more rapid fatal disease) would spread more successfully than the slower growing form, and it would then become the dominant strain.

Viruses which establish chronic infections and in effect play hide-and-seek with immune cells are discussed fully in Chapter 4.

Are viruses really a threat to the human race?

This question, which was posed in the introduction to this book, is a difficult one to answer. It is probably true that these days one of the biggest threats to mankind is from the use of viruses (and other micro-organisms) in biological warfare. This type of warfare is not by any means new, but now that we have access to state-of-the-art delivery devices which can disseminate any lethal micro-organism widely and efficiently, it is of international concern.

From time to time throughout history, peoples and governments around the world have used micro-organisms as an efficient and cost-effective weapon of mass destruction. Starting in a rather crude way, the Greeks and Romans deposited dead animals into their enemy's drinking water. Later, dead soldiers were added to this, and the technique was further refined in medieval times when the bodies of people who had died of infections were catapulted into besieged towns. In 1763, in the earliest recorded deliberate release of a virus, Sir Jeffrey Amherst, British Commander-in-Chief in North America, authorized the distribution of smallpox-contaminated blankets to native Americans who were harassing European settlers around the garrison at Fort Pitt in Pennsylvania.[10]

Moving to the twentieth century, our ability to produce large stocks of bacteria and viruses meant that biological warfare assumed global significance. In 1929, the Russians set up a biological warfare research station north of the Caspian sea, and this prompted Britain, Japan, the USA, and Canada to follow suit. Japan developed the most extensive programme and in the years leading up to and during the Second World War, they used human subjects in open-field trials to test out their lethal agents. Manufacture continued in some countries until the Biological and Toxic Weapons Convention came into effect in 1975. This certainly reduced the threat, but has not entirely eliminated it.

These days the main problem is likely to come from terrorist groups or megalomaniac dictators. For them, biological weapons have many advantages over their conventional counterparts, including being cheaper and relatively easier to prepare. Although new restrictions are in place, seed cultures of many dangerous micro-organisms can still be obtained from national collections, with no questions asked, and since making vaccines is a legitimate reason for growing microbes on a large scale, biological weapons factories can masquerade as vaccine production plants.

Micro-organisms are ideal for selective attacks on individuals or for targeting large metropolises. They can be smuggled through all traditional security devices, and only tiny volumes are needed to kill huge numbers of people. Furthermore, since they are invisible, odourless, tasteless, and have a delayed action, they can be released into the air without immediate detection. Finally, as we have no previous experience of this type of attack, they are bound to unleash such panic and psychological trauma that total confusion will reign.

Many different organisms have been tested for their potential as agents of biological warfare, including bacteria which cause TB, typhoid, plague, cholera, and gas gangrene. Candidate viruses include rotavirus which produces incapacitating vomiting and diarrhoea, and haemorrhagic fever viruses like Ebola which can be 90 per cent lethal. But the most effective biological weapons are now considered to be the bacteria causing anthrax and botulism, as well as the smallpox virus.

Because smallpox was eliminated from the natural environment in 1980, vaccination programmes have ceased and the world's population is once more susceptible. The virus causes a devastating and often lethal disease which spreads rapidly in large and crowded cities. But best of all from the point of view of the aggressor, the virus remains infectious for long periods, so that it could be packed into the war heads of guided missiles and sent to its destination still in a viable condition. The threat of a deliberate smallpox release is well recognized by governments, and stocks of vaccine have been retained for this eventuality. Preparations for 'Operation Desert Storm' during the Gulf War of 1990 included vaccination of US and British troops, but in reality it would be impossible to vaccinate an entire population in time to prevent a worldwide epidemic. However, biological weapons are primarily designed to destroy all vital activity but not necessarily to wipe out the whole human race. And smallpox, although it would certainly incapacitate, would probably not kill all those infected because of our inbuilt resistance, developed and strengthened during the centuries when the virus was rife.

In this chapter we have seen the ingenious ways in which viruses sneak into our bodies and hijack our cells. This is the beginning of their success. Then, by adapting rapidly to any changing situation they can outwit our immune systems and gain the advantage—in the short term at least. Such adaptability confers unpredictability, and this engenders fear and panic. Are viruses really beyond our control? Concern over what new and emerging threats may be in wait for us is constant and real, and this is the topic discussed in the next chapter.

Chapter 2

New viruses or old adversaries in new guises?

In 1994, an Australian racehorse trainer and 14 of his horses died of a mysterious illness.

'Killer virus has scientists baffled'

shrieked newspaper headlines around the world.[1]

The facts behind these headlines were indeed sensational. The story began in Brisbane when, at the beginning of September, a horse trainer, Vic Rail, bought a pregnant mare and stabled her with his other horses. Two days later the mare died of pneumonia and then both the trainer and a stable-hand who had nursed the sick horse became seriously ill. By the end of the month the trainer and 14 horses were dead; the stable-hand and seven other horses recovered.

This is the worst kind of nightmare for those with the job of investigating the disaster and stopping it spreading—in this case, Keith Murray and his team from Australia's top security animal health laboratory in Geelong near Melbourne. By the time they get to hear about it 11 horses are dead and their trainer is extremely ill.

It sounds like a case of poisoning but it could be an infection with the potential to spread like the plague throughout Australia. They have to act fast and they do. Tissue samples from the dead horses are flown in from Brisbane and scientists attempt to grow a microbe in cell culture. Veterinary surgeons, dressed up like spacemen for their own protection, try to infect healthy horses with the material, both by injection and by inhalation. Meanwhile, Australia's Consultative Committee on Exotic Animal Diseases imposes a category III alert—top of the emergency scale.

Just three days after receiving samples, scientists know what they are dealing with; a virus has grown in cell culture. One day later there is enough virus to see under the electron microscope and they discover typical paramyxovirus particles. When the team compare genetic material of the new virus with other paramyxovirus family members they find no perfect match, but the closest alignment is with morbilliviruses, a group which includes measles and canine distemper viruses. So, after just one week of intensive investigation, the team has a new and extremely dangerous measles-like virus which has killed 70 per cent of the horses and half the humans it has infected. They name it equine morbillivirus.[2]

By now all the experimentally infected horses have developed high fever and are having great difficulty in breathing, confirming the worrying suspicion that the virus can spread from one animal to another by inhalation. At this stage the sick horses are humanely killed and samples of their tissues, examined under the microscope, show the typical appearances of haemorrhagic fever and reveal why the virus is so lethal. It damages cells lining blood vessels causing them to leak fluid into surrounding tissues. In the lungs, where there is a vast network of blood vessels, there has been so much leakage that if they hadn't been killed, the animals would literally have drowned in their own fluid. Unfortunately, that is exactly what happened to the trainer and his 14 horses.

After identifying the cause of this new disease, the next priority for the team was to develop a blood test to diagnose it and then to identify animals infected in the past. This they achieved with amazing speed. Within a month they knew that no other horses or humans in the outbreak area had been infected. With this information the state of emergency ended and no more cases occurred. Veterinary surgeons then concentrated on solving the mystery of where the virus came from. They already knew that it was not from other horses in the area so they started screening every animal species in the suburb of Brisbane, where the first case—the pregnant mare—came from. But this time they were not so lucky; they were in for a long haul which only ended in May 1996. They screened more than 5000 blood samples from 46 animal species before their luck turned and they found what they were looking for. Eleven out of 55 fruit bats (flying foxes) tested had antibodies to the virus, strongly suggesting that they are the natural host for the virus, probably carrying it as a harmless infection. But these are not blood-sucking bats; they eat fruits, live high up in trees, and are usually shy of other animals. So how the first horse in this outbreak got infected remains a mystery.

This equine virus is by no means the only morbillivirus to emerge recently. While we are striving with considerable success to rid ourselves of measles, there have been several outbreaks in other species caused by related viruses. In 1988 wildlife experts were alerted when hundreds of dead harbour seals were washed ashore along the coasts of Northern Europe. The culprit was a new virus named phocine distemper virus, which eventually killed around 20 000 animals. More recently, over 1000 lions in the Serengeti National

Park in Tanzania died from a mystery disease. Tourists taking a balloon ride over the park were the first to notice sick and dying lions. This was quickly traced to canine distemper virus infection which is endemic among local dogs belonging to Maasai tribes' people. Dogs probably passed the virus to hyenas, foxes, and jackals which, in turn, infected the lions.

The dramatic story of equine morbillivirus illustrates how emerging infections usually occur: out of the blue, in a rural setting, and commonly acquired from animals (known as zoonoses). In this chapter we look at how and why these new virus infections emerge and spread, and the reasons behind their ever-increasing numbers in recent years.

Brand-new or second-hand?

The appearance of a new virus infection is not just a chance event. There are always rational answers to the questions 'why?', 'how?', 'when?', and 'where?' With every new infection, as with equine morbillivirus in Australian horses, answers to these questions are diligently sought by teams of scientific detectives dedicated to chasing up reports of strange new diseases, often in far-flung areas of the world. Once the culprit is found, scientists can map similarities between its genes and those of related viruses. Then they plot a family tree from which they can deduce the likely source of the 'new' virus.

Emerging infections are usually caused by viruses which are not strictly speaking 'new'. They have generally been around for millions of years but have changed their habits in some way. Most often they are viruses which naturally infect animals, coevolving with these primary hosts until they are relatively harmless to them.

Problems only arise when for some reason they cross a species barrier and colonize a new host. Occasionally it is a genetic mutation which allows this to happen, but this is much less common than was once thought. Mostly these days it is man's intrusion on the natural environment which is the all-important key.

There are many factors which interact to produce the ideal circumstances for a virus to flourish. In adapting to our natural surroundings over time, the human lifestyle has changed from nomadic to agricultural to urban, and viruses have exploited these changes to their maximum advantage. Now, more than ever before, we are affecting our natural environment and, increasingly, the local flora and fauna and the viruses they carry. Colonization of new territories in Africa, deforestation for development of new farming land in South America, intensive factory farming in North America and Europe, domestication of exotic animals for pets, large irrigation schemes, wars, and global warming are just a few of the ways we make ourselves felt. All impact on the ability of viruses to adapt to changing environments, inducing ecological changes which place people in contact with new reservoirs of infection. And this is no longer just a local problem; rapid air travel can carry infection from one community to a new country or continent in just one leap.

The following accounts of recent disease outbreaks illustrate some of the factors involved in the appearance of 'new' virus infections.

Haemorrhagic fever

In addition to morbillivirus infections, several other viruses cause forms of haemorrhagic fever and are among the most lethal diseases known to man. These include the deadly Ebola, Lassa, and Hanta viruses, as well as yellow fever and dengue fever viruses (all discussed in Chapter 1). Infection causes fever; head, muscle, and abdominal aches; but, as the name suggests, bleeding is the hall-mark. Bleeding into skin can produce bruising and a characteristic purple-coloured rash, while internal bleeding destroys vital organs. Haemorrhagic fevers are not new; they have attacked man before. The Chinese have written records of outbreaks dating back over 1000 years.

A temporary fluctuation in climate was responsible for an out-break of Hanta virus infection in the USA in 1993. One day a pre-viously healthy young couple were rushed to hospital in New Mexico with severe flu-like symptoms—fever, headache, cough, and difficulty in breathing. Both died rapidly and doctors were puzzled. They made inquiries locally and found that patients with similar symptoms had been seen in other hospitals in New Mexico and in neighbouring Arizona and Colorado (Fig. 2.1). Clearly there was a new and terrifying disease around so the State Health Department and The National Center for Disease Control and Prevention in Atlanta took over the investigation. By the end of 1995 a total of 119 people had suffered from the mystery disease and 58 had died.

Scientists identified a new type of Hanta virus which they called 'Sin nombre' (Spanish for 'without name'). The disease it causes is now called Hanta virus pulmonary syndrome. They screened several hundred people at risk of infection, including patient con-tacts, family members, and their carers, but found no evidence of spread between humans. A search among local wild animals for the natural host finally identified the deer mouse (a common North American rodent). About 20 per cent of these mice in New Mexico,

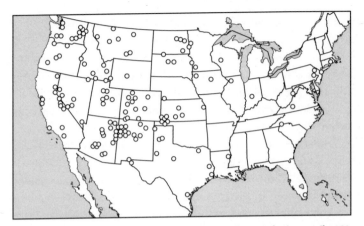

Fig. 2.1 Map of the USA showing cases of Hanta virus infection until 1998. (Based on map from 'Hanta virus: an emerging threat to human health?' by D. Bonn. *The Lancet*, **vol. 352**, p.886, 1998, copyright The Lancet Ltd.)

Arizona, and Colorado carry the virus, but this finding alone does not explain the 1993 outbreak since humans and deer mice have happily coexisted in the South West States of the USA for centuries.

Hanta virus passes easily from one generation of mice to the next in nests, where young deer mice are in close contact with older infected mice. They do not become ill, but once infected they excrete the virus in saliva, faeces, and urine for long periods.

The winter of 1992–3 was exceptionally mild, allowing more older mice than usual to survive. The following spring was particularly wet, producing an abundance of wild pinon nuts which deer mice like to eat. Consequently the deer mouse underwent a population explosion, spread more widely, and came into closer contact with humans in rural areas. Humans catch Hanta virus by inhaling infected material—that is dust contaminated with mouse urine or faeces. Most people who caught it in this outbreak either lived or worked in mouse-infested cabins, barns, or outhouses.

Since finding this new Hanta virus, scientists have identified the same virus in around 15 per cent of samples stored from people who died previously in South-West USA of unexplained flu-like illnesses. So the virus must have jumped species and infected man before when conditions were favourable.

Similar outbreaks of haemorrhagic fever caused by related Hanta viruses and carried by local rodents occur regularly in Scandinavia and Russia. These are less severe because they lack the acute lung symptoms seen in the US cases. But disease incidence similarly rises and falls according to the rodent population density and correlates with natural variations in climate. Other outbreaks have resulted directly from disturbance of the local environment. During the Korean War, for example, 2000 UN troops were infected, and more recently there has been an outbreak in the war zone in Bosnia. Here, military camps with large food stores probably attracted the natural virus-carrying local rodent population and primitive overcrowded living conditions gave the virus its chance to infect man.

In the 1940s, changes in agricultural practices were responsible for the appearance of Argentine haemorrhagic fever caused by Junin virus. This virus is carried by rodents which eat maize. As the population of Argentina grew, large areas of pampas were cleared and replaced by maize. Numbers of maize-eating rodents consequently escalated and the incidence of haemorrhagic fever among the farmers increased in parallel. Nowadays, it is operators of mechanical harvesters rather than farmers who are at risk because they inhale the clouds of dust contaminated with rodent excreta that are generated by the machines.

Although Hanta viruses often cause fatal disease, doctors have been comforted by the fact that their impact is limited because they do not spread from one human being to another. Each person must be infected individually from the natural host—a factor which fortunately prevents large epidemics. But for the first time, in 1997, during an outbreak in southern Argentina, five doctors who looked after patients with Hanta virus caught the infection. This is strong evidence that the virus spread directly from patient to doctor without an intervening mouse host.[3] If true, this alarming finding means that Hanta virus can produce the lethal epidemic pattern of other haemorrhagic fever viruses like Ebola and Lassa fever, where direct person-to-person spread is key to their explosive effect. These viruses are transmitted by contact with body fluids, and since they can cause catastrophic bleeding and diarrhoea, the infection often spreads to family members, carers, and laboratory staff before the correct diagnosis is made.

The first recorded Ebola outbreak in Yambuku, North Zaire, in 1976 (outlined in Chapter 1) was typical. It appeared in a remote area, without warning, spread rapidly from a single case, and caused fear and panic. Outbreaks like this have continued over the intervening 20 years, one of the latest being in gold-mining camps in the remote rain forests of Gabon, where again the death rate was extremely high. In each of these epidemics the virus has infected humans from an unknown (presumed animal) reservoir in the rain forests of Africa, but despite exhaustive trapping and testing of wildlife in the areas, no infected animals have been found. Until the primary host for Ebola virus is identified it is impossible to take precautions, so new outbreaks will surely occur.

The first recorded victim of Lassa fever was an American nurse working at a mission station in the town of Lassa, North-East

Nigeria, in 1969. She developed a severe flu-like illness and was eventually flown to the Evangel Hospital in Jos where she died. The nurse who looked after her caught the virus and died 11 days later. Next, the head nurse at the Evangel Hospital fell sick and was flown to the USA (in the first-class cabin of a commercial aircraft, with no special precautions taken) where she slowly recovered. Virologists at Yale Arborvirus Unit set to work and isolated a virus from her blood which they called 'Lassa'. While working on this one of the scientists became ill and was only saved by an infusion of plasma from the nurse who had already recovered. Finally, five months later a laboratory technician from the Yale Unit came down with Lassa fever and died—although he had never worked directly with the virus. This chain of events, coupled with several later, more devastating outbreaks in West Africa, gave Lassa virus its well-earned awesome reputation.

Lassa virus is carried by African brown rats which get infected at birth and excrete the virus freely throughout their lives. These rats are common in and around towns and villages in sub-Saharan Africa and during the rainy season they find shelter in buildings, so coming into closer contact with humans. In these areas infection is quite common but not particularly lethal—although the virus is still responsible for around 5000 deaths a year.

In 1989, the city of Chicago had a lucky escape. A man with flu-like symptoms at the height of a winter flu epidemic was not surprisingly assumed to have a severe dose of flu. His case raised no suspicions for eight days, by which time he was seriously ill. Only then was he questioned closely and his story emerged. He had attended his mother's funeral in Nigeria two weeks previously, and ten days after that his father died. The information that both parents had illnesses similar to his own set alarm bells ringing; he had Lassa fever. He died shortly after the diagnosis was made, but during his illness, when he was presumably highly infectious, he had been in contact with 102 people. All were traced but fortunately none developed the disease.

Surprisingly, the frighteningly high mortality rate among those suffering from Ebola and Lassa fever means that these viruses do not cause widespread epidemics. Because they rapidly incapacitate most of the people that they infect, and kill within a week or two, they have to rely on spreading more rapidly than this if they are to survive in humans. This is unlikely to be a practical proposition for viruses which spread by direct contact, so they typically cause explosive but localized epidemics. If they were capable of spread by inhalation like measles or flu viruses, or had a long, silent incubation period like HIV, it might be quite a different story.

Bats in the belfry

Rabies is an ancient and highly lethal virus (Fig. 2.2). It manages to survive in today's world despite having to rely on animals biting each other for transmission from one host to another (Fig. 2.3). Rabies skilfully drives its victims into a mad frenzy which lasts several days, during which time the animal's saliva is loaded with virus. From an infected bite the virus homes in on nerves in the skin and begins a long journey up the nerve fibre to the brain. While travelling it causes no symptoms at all—the wound heals and all seems well. But eventually, anything from seven days to several years later (a period determined by the distance from the original bite to the central nervous system), the virus reaches the brain. Here it causes inflammation (encephalitis) which induces bizarre behavioural changes. Wild wolves, which are usually solitary, shy animals,

A

DECLARATION

OF SVCH GREIVOVS

accidents as commonly follow
the biting of mad Dogges,
together with the cure
thereof,

BY
THOMAS SPACKMAN
Doctor *of* Physick.

LONDON
Printed for *Iohn Bill* 1613.

Fig. 2.2 Title page from Sir Thomas Spackman's treatise on rabies of 1613.
(Courtesy of Royal Society of Archives.)

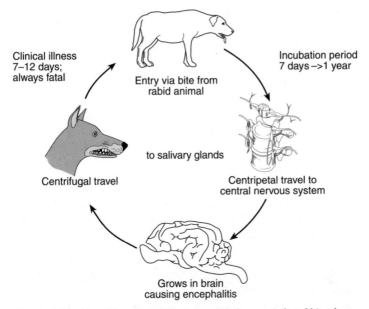

Fig. 2.3 Rabies virus life cycle. Rabies virus travels from an infected bite along nerves to the brain where it grows and causes encephalitis. It then moves to organs including salivary glands where its presence in saliva allows its transmission to another host.

seek out the company of other wolves or approach human habitation. Humans with rabies become hyperactive, delirious, and often violent. Severe muscle spasms, particularly induced by drinking, cause the classical hydrophobia. This behaviour is interspersed with lucid intervals when recovery seems possible, but inevitably after a week or so sufferers sink into a coma and die. While all this is going on, the virus is on the move again. From the brain it travels back along nerves, this time to tissues including salivary glands. Here it grows and releases masses of virus into the saliva of the infected animal which is raving mad and ready to bite anything that approaches it.

Living on an island continent, Australians have nothing to fear from the deadly rabies virus—or so they thought. But in November 1996, a healthy 39-year-old woman suddenly noticed numbness and tingling in her arms and legs. She then developed fever and vomiting, and by the time she was admitted to the Royal Brisbane Hospital she had full-blown encephalitis.[4] She rapidly sank into a coma and despite all efforts to save her, she died. Doctors ascertained that a few weeks before her fatal illness the woman had been scratched by a sick fruit bat she was caring for. Australian fruit bats may look cute and friendly, but only two years before the women was scratched they were identified as the carrier of equine morbillivirus—so the woman's story alerted the virus hunters. They found a new type of lyssavirus (the group to which rabies belongs) which is 92 per cent identical to rabies, and probably just as lethal. Luckily this close similarity means that rabies vaccine works for its cousin as well, so Australian bat handlers and fruit pickers can be offered protection against this new killer.

Acquired immunodeficiency syndrome (AIDS)

AIDS appeared to come from nowhere when it exploded like a bombshell in the early 1980s. Just as we were lulled into a false sense of security by our success in controlling other infectious diseases, a new plague hit us. The press had a field day; here was a disease associated with gay men, promiscuity, drug abuse, and prostitution—all the ingredients for a media sensation. Journalists vied with each other for the most lurid stories and AIDS rapidly became 'the gay plague', fuelling prejudices and causing panic. Not since syphilis was rife has such a sense of fear and guilt been engendered by an infectious disease. Many headlines were judgmental and discriminating:[5]

'Place AIDS victims in quarantine';
'Blood on their hands'; 'Exterminate gays'.

In the early days, discrimination even extended to young children like Eve Grafhorst from New South Wales in Australia, who caught the virus from infected blood. When this leaked out she was banned from the local kindergarten for fear that she might infect the other children:

'AIDS—Little Eve banned from a play centre'.

When she returned to school several parents removed their children:

'Forty children withdrawn as AIDS toddler returns'.

She was accused of biting another child and expelled; her family was hounded and evicted from their house. Finally they sought refuge in New Zealand.

The first cases of AIDS were reported in the medical journals in 1981. *The Morbidity and Mortality Weekly Report* from the Center for Disease Control in Atlanta, recorded, 'five young men, all active homosexuals, were treated for biopsy-confirmed *Pneumocystis carinii* pneumonia in three different hospitals in Los Angeles, California'.[6] Just one month later came the report of 26 cases of Kaposi's sarcoma—until then an extremely rare tumour in the USA—again all in previously healthy, gay men.[7] The appearance of these two diseases (generally only seen in people with severely suppressed immune systems) in previously fit and healthy, young, gay men had doctors baffled and prompted a countrywide study. Very soon AIDS cases were popping up in places as far apart as Haiti, Europe, and Australia and the race was on to find the cause.

The primate connection

Human immunodeficiency virus (HIV) was isolated in 1983 from lymph gland tissue from a French gay man,[8] and now there is overwhelming evidence to show that HIV causes AIDS (discussed in Chapter 4). But where did the virus come from?

Like most emerging infections, HIV crossed the species barrier from animal to man, but the particular animal involved has been difficult to track down and all along there have been plenty of scare stories—a man-made virus released by the US Government; a contaminant of polio vaccine; a virus from an as yet undiscovered animal lurking in the African jungle.

There are two types of HIV—1 and 2—and most scientists would now agree that the epicentres for both epidemics were somewhere in Africa. Searching through banks of stored blood samples, scientists found HIV-1 infections of humans in Central Africa as far back as the 1950s, and HIV-2 in West Africa since the 1960s. The low level of infection found at that time was not enough to cause an epidemic but since then, whereas HIV-2 has remained localized to West Africa, HIV-1 has spread globally. Blood samples taken during the 1976 Ebola epidemic in Zaire show that less than one in a hundred people were infected with HIV-1, and repeat testing 10 years later showed no increase. However, jump another 10 years to 1996, and infection in the same region had reached epidemic proportions.

Many African monkey species carry viruses known as simian immunodeficiency viruses, or SIVs, which are similar to HIVs. We have known for some time that HIV-2 is so closely related to SIV of the sooty mangabey monkey that this is its likely origin. This animal is sometimes hunted for food or kept as a pet in West Africa, and over the years humans have probably occasionally picked up the virus though infected bites or cuts. These few scattered cases attracted no particular attention, but eventually the virus established a sufficient toe-hold to spread successfully between humans.

Chimpanzees have long been the suspected source of HIV-1, but until recently the story was confusing. Chimps live in 21 different African countries, and whereas some chimp SIVs closely resemble HIV-1, others are quite different. Beatrice Hahn and her co-workers from the University of Alabama, Birmingham, USA have compared SIVs from four different captive chimps and come up with the probable answer.[9] The story began in 1995 when Hahn was offered samples from a chimp called Marilyn who tested positive for antibodies to HIV-1. Marilyn was not sick at the time the samples were taken, but died giving birth shortly afterwards. Hahn found that SIV from Marilyn closely matched HIV-1, as did SIVs from two chimps caught earlier in Gabon. However, SIV isolated from Noah, a Zairean chimp, was quite different.

Chimp subspecies have evolved in isolated geographical territories defined by river valleys, and they do not interbreed. When Hahn analysed the four chimps' DNA she found that they belonged to two different subspecies. All three chimps with HIV-1-like SIVs were subspecies *Pan troglodytes troglodytes*, whereas Noah from Zaire was *Pan troglodytes schweinfurthii*. The territory of *Pan troglodytes troglodytes* maps precisely to the area of Africa where the first human HIV infections were recorded, so Hahn is convinced that this SIV is the direct ancestor of HIV-1. She speculates that the virus must have transferred to humans on more than one occasion during hunting, slaughtering, and dismembering of chimps for 'bush meat'.

Virus hopping

HIV infects CD4 positive T lymphocytes (see page 30) in the blood and can be spread directly to uninfected people by blood transfusion. By this route the virus immediately finds itself in contact with new CD4 T cells that are ready and waiting to be infected. But with millions of viruses in every millilitre of blood, any contamination, be it from an infected needle or during the birth of a baby, may be enough to sow the seed of a new infection.

One of the tragedies of the HIV epidemic was the way haemophiliacs were infected when HIV found its way from donated blood into human clotting factor VIII (required by haemophiliacs to prevent bleeding). During the first decade of the AIDS epidemic over 1000 haemophiliacs in the UK contracted HIV, and the death rate among haemophiliacs as a whole rose tenfold. However, as soon as the problem was recognized it was relatively quick and easy to control, and now HIV testing of blood and blood products is routine in almost every country.

Another important way in which the epidemic spread was by intravenous drug abuse, and this has not been so easy to control, mainly because it is illegal. In the early 1980s, drug users were not supplied with clean needles and syringes so old needles were shared and reused without a thought for their sterility. As a result, HIV, along with other blood-borne viruses, passed freely from the blood of one needle user to another. A particular problem came to light in prisons where up to half the inmates are intravenous drug users. Prisoners depended on injecting equipment being smuggled in by visitors, so obviously needles got reused many times. Although HIV is most likely to spread in prisons when the incidence in the surrounding area is high, this is not always the case. An outbreak among prisoners in Thailand occurred when the prevalence in the local community was still low, and a Royal amnesty in 1988, by inadvertently releasing large numbers of HIV-positive intravenous drug users into the community, probably accentuated the developing epidemic in the population at large.

But HIV could not succeed in causing a worldwide epidemic by injection alone. The virus is primarily spread by sexual intercourse, be it heterosexual or homosexual contact. Both male and female genital secretions contain virus, although transfer is more efficient from male to female and from active to passive gay partner than vice versa. This is because infection is more likely when virus-contaminated semen is deposited directly onto mucosal surfaces. Although the virus can cross the intact lining of male or female genital tracts, sexual practices which predispose to injury or ulceration—including anal sex, dry sex (where desiccants inserted into the vagina dry up lubricating secretions), and genital ulceration caused by other sexually transmitted diseases (STDs)—increase the risk of infection. Conversely, practices which prevent injury, such as circumcision (which results in thickening of the skin on the tip of the penis) and condom use, protect from infection.

Since gonorrhoea and syphilis both cause genital ulceration and inflammation, which increases the number of blood cells (including CD4 T cells) in genital secretions, it seems reasonable to assume that these infections in either partner increase the risk of HIV infection. On the one hand there would be more virus to spread to a sexual partner, and virus from an infected partner would gain entry more easily through the raw, ulcerated surface on the other. But proof of this is difficult because the type of behaviour which puts people at high risk is the same for both HIV and other STDs. However, the results of two studies carried out in the early 1990s lend support to this notion. In the first, Kevin De Cock, an epidemiologist from Centre for Disease

Control, Atlanta, USA, working on the Ivory Coast, West Africa, showed that commercial sex workers with genital ulcers transmitted HIV to their partners more often than those without. Also, successful treatment of genital ulcers reduced genital shedding of HIV.[10] The second study was masterminded by Richard Hayes, an epidemiologist from the London School of Hygiene and Tropical Medicine working in Mwanza, a rural area of Tanzania where STDs are common and poorly treated.[11] Hayes chose 12 similar villages; in six he implemented modern treatment for STDs whereas in the other six, STD sufferers received the standard, rather ineffective, local treatment. After just two years the incidence of HIV in both men and women in the six villages where no modern treatment was available was twice that of the villages where effective treatment had been given. So both studies show that HIV transmission is enhanced by concurrent STDs.

Going global

Human migration and industrialization go hand in hand, and in Africa this has played a key role in the spread of HIV. Increasingly people are moving from rural areas into cities looking for work, so that by the year 2025 the United Nations estimate that over 60 per cent of the population of the developing world will live in cities. Initially, HIV moved with this migrant population along the highways to become a city dweller. Once established in cities where there was more potential for spread between individuals, the infected population grew rapidly. People with multiple sexual partners, like female commercial sex workers, became infected early on and acted as a focus for dissemination of the virus. In townships of migrant workers, where men live for long periods at a distance from their families, commercial sex flourishes. South African mining towns are one such example. The large numbers of men and few women are a recipe for commercial sex and multiple sexual partners—the milieu in which HIV thrives.

Once infected, city workers carried the virus back to their villages, infected their partners, and through them often their future offspring as well. Studies in Africa in the late 1980s found more infected people in villages with easy access to large towns or cities, like those situated on highways or served by a regular bus service, than in more remote and isolated villages. In one area of Uganda, banana traders cycling daily between market towns and villages were identified as carriers of the infection.

So, silently, HIV infection reached frighteningly high levels before the problem was fully appreciated. In the space of 20 years it mushroomed from being an unrecognized, isolated, rural infection of central Africa, into an epidemic infecting 30 million people worldwide (Fig. 2.4). HIV has already caused over 11 million deaths and now outranks malaria as the leading killer in Africa. Every six seconds, someone in the world becomes infected with HIV; there is hardly a single country which is not affected by this virus. How has HIV achieved this remarkable feat?

All over the world, sexual networks, like ever-growing spider's webs, spread the disease. A small cluster of cases in Belgium amply demonstrates this relentless pattern.[12] One heterosexual man had many sexual partners concurrently, and when he was diagnosed as HIV positive, 19 of his most recent partners were traced. Eighteen were willing to be tested for HIV and 11 had been infected by him. Each of these women had one or more other sexual partner, who also had other partners. And so the network grew. This, multiplied millions of times over, often with the source of infection being a commercial sex worker, is the way that HIV has achieved its global spread (Fig. 2.5).

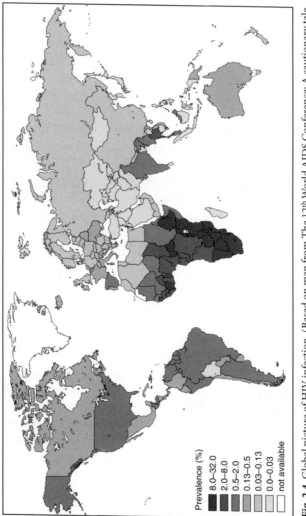

Fig. 2.4 Global picture of HIV infection. (Based on map from The 12th World AIDS Conference: A cautionary tale by R. Horton. *The Lancet. The Lancet*, **vol. 352**, p.122, 1998, copyright The Lancet Ltd.)

Fig. 2.5 HIV poster. Reproduced with permission of Texas Department of Health.

The Edinburgh experience

Doctors in Edinburgh were first alerted to the AIDS epidemic on their doorstep in 1984 when 44 per cent of a group of haemophiliacs tested HIV positive.[13] They had only received locally produced factor VIII, so the hunt was on to find the HIV-positive donors. The search identified nearly half of Edinburgh's intravenous drug users as positive, compared to just 4.5 per cent in Glasgow and 10 per cent in England and Wales. How could this have happened?

Intravenous drug use became a problem in Edinburgh in the late 1970s and had already peaked in 1983 before doctors even knew it existed. But the police knew and treated it aggressively. They discouraged surgical supply shops and pharmacists from supplying users with syringes and needles, and when they seized drugs from users they also removed their injecting equipment. The police used this evidence of illicit drug use to persuade users to expose their suppliers. In response, drug pushers forced their visiting clients to inject on site and leave the premises free of incriminating evidence. Combined with the severely limited supply of equipment, this led to users gathering in a few large 'shooting galleries' and sharing one set of equipment between anything up to 40 people.

Most intravenous drug users in Edinburgh were young, local people who never left the city. A few, however, gravitated to southern Europe where heroin was cheaper and easier to find, but also, unfortunately, where HIV had been rife since 1979. One particular drug user returned to Edinburgh from Spain in early 1983 and almost immediately developed jaundice due to hepatitis B virus infection. He stayed in Edinburgh for between 6 and 12 months, sharing equipment with many other users, and then disappeared. He resurfaced in 1987 when he tested HIV positive. Looking back at his blood samples stored from the time of his hepatitis B infection, doctors found that he had in fact become HIV positive in January 1983. He is the first-known HIV-positive case among intravenous drug users in Edinburgh and many of the people who shared his equipment became positive around August 1983. From this small beginning the final toll in Edinburgh was over 1000 cases of HIV out of a total of around 2000 intravenous drug users in the city—the most HIV-positive people per head of population anywhere in the UK. The majority were young and, unusually for HIV, a third were female. The only fortunate thing about the whole episode was that infections were identified early and could be followed and treated aggressively. Despite this, 70 per cent had developed HIV-related illnesses by 1993.

The Edinburgh epidemic was unusual in the West where the HIV epidemic began, and has largely remained, in the gay population. This has been devastating for the gay community and is difficult to explain. Sometime in the 1970s American gay men visiting African cities must have picked up HIV which was by then silently spreading. Infection first took hold in Los Angeles, San Francisco, and New York where a subgroup of gay men had an extraordinary lifestyle. Bath houses were a hot bed of casual sex, with some men reporting over 40 different partners every week. Anal sex, combined with frequent STDs, encouraged the rapid spread of the virus in this particular group which to date has accounted for most of the 600 000 AIDS-related deaths in the USA and Europe. While infection rates in gay men in the major American cities have stabilized (probably due to saturation of those with the appropriate lifestyle and/or a change to safer practices), the epidemic is now radiating out to smaller cities where infection rates are still rising.

Bisexual men are the link between homosexual and heterosexual populations and in the 1980s it was predicted that through them an HIV epidemic in heterosexuals was imminent in the West. To a certain extent this is happening; the rate of spread of HIV between heterosexuals is increasing rapidly and fast catching up with homosexual transmission. However, since we are now aware of the problem, this epidemic is unlikely to reach the dramatic proportions of that seen in gay men in the 1980s and in Africa today.

HIV as a human exterminator

The threat that HIV could wipe out the human race now seems far fetched and alarmist, but at the beginning of the epidemic, mathematical modelling of HIV spread suggested that within a relatively short time a large proportion of the world's population could be infected.[14] Clearly this prediction failed to take into account several key factors.

As soon as HIV was discovered, an unprecedented campaign of public awareness and education combined with public health measures were rapidly implemented. These included needle exchange systems to reduce sharing by drug abusers, free condom distribution from STD and family planning clinics, and codes of practice for health care workers to prevent infection by accident or injury. It is difficult to measure the effect of these initiatives on the overall picture of the epidemic, but they must have played an important part in restricting the spread of HIV in the West. And now, at last, new drug combinations (detailed in Chapter 6) can control infection and prevent spread, so that for the first time, the death rate among HIV infected people is falling.

However, while there is some optimism in the West, the situation elsewhere is bleak. In most developing countries, where a large percentage of the population are young and sexually active, HIV is still spreading rapidly. India, with four million HIV positive, now has the highest number of HIV-infected people in the world. Surrounding countries, including China with its massive population, are poised at the forefront of the epidemic. This is where effective measures to stop the advance of HIV and drugs to treat those already infected are urgently needed, but the Indian Government cannot afford either. The annual cost of a course of treatment, at around £15 000, balanced against the annual health budget of between £1 and £2 per person in many developing countries, obviously puts treatment out of the question. AIDS has fast become a disease of the poor in poor countries.

Doctors have noticed a few people with high-risk behaviour, like gay men and female sex workers with regular HIV-infected partners, who always test negative for HIV. These people, in common with around 1 per cent of the world's population, carry a genetic mutation which prevents HIV from infecting their cells. This remarkable finding underlines the genetic diversity of the human race and indicates that even in the most devastating epidemic, like a hypothetical HIV spread by inhalation, some people would probably survive.

Mad cows and Englishmen

It was in 1986 that UK farmers first noticed a strange new disease spreading through their cattle herds. Now recognized as bovine spongiform encephalitis, BSE was rapidly dubbed 'mad cow disease' because of the brain degeneration and fatal paralysis it causes. Although BSE is not caused by a classical virus, it is included in this book because, like a virus, the agent is infectious and can be transmitted in a cell-free form. The epidemic and the search for its cause are described here, whereas the nature of the BSE agent is discussed in Chapter 4.

BSE struck over half the dairy herds in the UK as well as zoo animals (the oryx, kudu, eland, puma, and cheetah), domestic cats, and, most recently, humans. The epidemic peaked in 1993 when over 160 000 cattle contracted BSE. Indeed the total number of infected animals is probably nearer 900 000, since the long incubation period of BSE—around five years—means that many infected animals were culled for beef before developing symptoms.

Why feed cattle to cattle?
Scientists set to work to discover the cause of this new disease and soon identified a particular protein-rich dietary supplement, meat and bone meal, which had been fed to animals on all affected farms. But how could this be the cause when it had been fed to animals for years without ill effect?

The story began in the early 1980s when the world price of tallow dropped. Tallow is a fat used in the cosmetic industry and it is a by-product of the rendering process in which carcases of cattle and sheep are boiled down to make meat and bone meal. With the drop in profits from tallow, economies were made in the rendering process. Organic solvents were no longer used to extract fat, and since it was therefore not necessary to recover these solvents for reuse, the time the brew was held at high temperature was cut. These apparently innocuous alterations were enough to allow the infectious BSE agent in the carcases to survive.

Although the source of BSE was quickly discovered, the origin of the infectious agent was not so easy to uncover. Scrapie, a BSE-like disease of sheep, has been around in UK flocks for several centuries, so inevitably infected sheep carcases were rendered for meat and bone meal and fed to cattle. Consequently the scrapie agent could have crossed the species barrier and infected cows, but equally, cattle may have their own BSE agent which under natural circumstances so rarely causes disease that it passed unnoticed until the practice of feeding cattle meat from their own species. Either of these two alternatives, combined with the changes in the rendering process, could have allowed the agent to recycle within their food chain but at this stage it is probably impossible to distinguish between them.

By 1990, as public disquiet over the safety of beef mounted, the British Government did its best to allay national and international fears. The Minister of Agriculture, John Gummer, in a now famous news clip, tried to persuade his daughter Cordelia to eat a hamburger made from British beef. When she refused it, he ate it himself. Headline writers had a field day:

'Gummer beefs up campaign';

'Please swallow this burger, darling and help my career';

'Beef is perfectly safe, says Gummer' (Fig. 2.6).

"I AM ABSOLUTELY CONFIDENT THERE IS NO RISK"

Fig. 2.6 Cartoon from *The Independent* depicting John Gummer's attempts to allay public fears about BSE.

All this media hype only served to increase alarm and the inevitable happened:

> 'Beef sales plummet as boycott spreads';
> 'Choose something else for Sunday lunch';
> 'Beef off menu for 1m children'.

A few still believed that British beef was safe:

> 'Stop beefing.
> Has the whole country gone mad cow mad?';
> 'Mad about cow disease'.

The British Government maintained that this was the case but their assurances were to rebound badly when it later became clear that BSE had spread to humans and that therefore British beef had certainly not been safe to eat.

To be fair, at the beginning of the epidemic there were very few facts to go on—BSE has a very long incubation period and is caused by an unconventional agent which we knew very little about, and there was no laboratory test to diagnose infection before it became clinically obvious. Consequently there was no way of knowing how many cows were infected or whether BSE could spread between cows in a herd. Vital experiments set up to find out which animal tissues and secretions (such as milk) were infectious took up to 18 months to provide an answer. So, to start with, the committee of experts—SEAC (the Spongiform Encephalitis Advisory Committee)—had to extrapolate information from the known facts about scrapie in sheep.

A ban on feeding ruminant-derived protein to cattle, implemented in 1988 should, in theory, have put a stop to the epidemic. Indeed there was a sharp fall in the number of BSE cases, but still over 26 000 cattle which later developed BSE were born after this feed ban was brought in. It is now clear that most of these infections were due to farmers continuing to use contaminated animal feed which was still available, supposedly for feeding pigs and poultry.

At this stage the question on everyone's lips was: could BSE infect humans? It seemed only a remote possibility because we had lived with scrapie in our sheep for centuries, and had consumed sheep meat and offal, with no problems. On the other hand, if BSE was caused by the scrapie agent in cattle feed, then it had succeeded in crossing the species barrier between sheep and cattle. Why not make another jump from cattle to man? To be on the safe side, in 1989 the British Government introduced the 'specified offal ban' which removed potentially infectious material (brain, spinal cord, thymus, tonsil, intestine, spleen) from the human food-chain. But unfortunately, like the cattle feed ban, this was not rigorously enforced. Before the 1989 ban there were around 8000 cases of BSE, with many more cows incubating the disease without symptoms. All of these cows entered the human food-chain and thus British people were exposed to the BSE agent in their diet for several years before 1989.

After the ban, beef was certainly safer to eat than at any time since the early 1980s, but despite this some contaminated beef still found its way on to the dinner plate. In 1995, the EU put many British dairy farmers out of business by banning sales of their beef abroad. In response all British cattle over the age of 18 months, which might have been incubating BSE, were culled and removed from the human food-chain.

This whole episode cost the British taxpayer a total of four billion pounds. When the BSE agent was found in nervous tissues inside the spine and in the bone marrow of infected cows, the British Government banned the sale of beef on the bone 'on a strictly precautionary basis'. This controversial ruling, apparently implemented to protect the public from infection, rather undermined the political assurance that British beef was as safe to eat as it was before the BSE crisis began. However, it was mostly the consumption of meat from infected animals prior to 1989 which caused the human disease we now know as new variant CJD.

Transmissible spongiform encephalopathies (TSEs)

BSE belongs to a group of diseases known as TSEs, and although they only came to public notice in 1986 when the BSE epidemic hit British cattle, they are not new. Scrapie in sheep and CJD and kuru in man have been around for many years and have many parallels with BSE and new variant CJD.

Scrapie (so called because it causes intense itching which makes affected animals rub up against posts and fences for relief) was first noticed in East Anglian sheep in 1732. It still infects flocks of sheep and goats in several countries, and is particularly common in northern England, Scotland, and Iceland. Scrapie is naturally passed from an infected ewe directly to her lambs, but adult sheep can also pick up the infection by grazing on pastures where placental tissues remain after lambing. So the scrapie agent, like BSE, at some stage entered the food-chain, and this is probably why the disease is relatively common.

In 1935 an epidemic of scrapie in Scotland killed over 1000 sheep. The outbreak was traced to a contaminated batch of vaccine given to protect the sheep against 'louping-ill'. Since only sheep vaccinated with that particular batch developed scrapie, this was the first suggestion that TSEs are infectious. After an incubation period of one to five years, infected animals start with the characteristic intense and persistent itching, then develop an unsteady, staggering gait, and die within a few months. These symptoms are caused by degeneration of the brain which may look normal to the naked eye but under the microscope, displays the characteristic appearance of TSE—the grey matter is full of tiny holes giving a spongy texture, and there are patchy deposits of an insoluble material called 'scrapie-associated fibrils' which destroys normal brain tissues.

H.D. Creutzfelt and A. Jakob both spotted the human equivalent of scrapie, Creutzfelt–Jakob disease (now called CJD), independently, in Germany, in the early 1920s. The disease occurs worldwide, affecting about one person in a million each year. Sufferers are usually elderly and die from dementia within six months. How CJD spreads naturally is a mystery, but it can certainly be transmitted like an infection because it has occasionally been transferred in contaminated human material taken unwittingly from people incubating the disease. For example, CJD has been passed to recipients of corneal transplants used to treat blindness, by grafting membranes during brain surgery, and by injections of growth hormone prepared from human brains. There are several other human TSEs which are all very rare, but by far the most interesting and well-known is kuru.

Daniel Carleton Gajdusek, an American paediatrician, who eventually won a Nobel prize in 1976 for his work on kuru, went to Australia in 1955 to work with the immunologist Sir Macfarlane Burnet at the Walter and Eliza Hall Institute in Melbourne. As it happened, Burnet had just returned from Papua New Guinea where he had seen people of the remote Fore-speaking tribe of the Eastern highlands who were suffering from a fatal neurological disease called kuru. In Fore, the word 'kuru' means 'to be afraid' or 'to shiver', a name which portrays the unsteadiness and tremor which develop as the disease takes hold. Apparently kuru was a new disease which mainly affected women and children, and was the commonest cause of death amongst them at the time.

Gajdusek was intrigued by the description and decided to go and see for himself. He spent 10 months living with the tribe studying their habits and customs, and concluded that kuru was probably a genetic disease.[15] He returned to the USA in 1957 where he continued to work on kuru. He was particularly struck by its similarity to Parkinson's disease because of the tremor characteristic of both illnesses, but after the similarities between kuru and scrapie were pointed out in a letter in the *The Lancet*,[16] his research changed tack. Because scrapie is infectious, he tried transmitting kuru to animals by inoculating chimpanzees with brain tissue from kuru victims, and 20 months later the first chimpanzee developed the symptoms of kuru.[17] From here, he passed the disease on to further chimps and to other types of monkeys, so proving that kuru was infectious. The cause, he suggested, was the 'kuru virus' which was spread among the Fore people by the ritualistic cannibalism which they practised. During these ceremonies it was customary for women and children to eat the brains of dead relatives, and consequently they were the ones who became infected and exhibited the highest incidence of kuru.

Gajdusek was probably right about this because since cannibalism was stopped in the late 1950s, kuru has slowly disappeared. However, the disease is not caused by a classic virus and the reason for kuru's appearance in a single tribe remains a mystery. In a similar way to BSE 40 years later, it may have been the result of eating a brain from someone incubating the rare natural form of CJD, with the subsequent entrance of the infection into the (somewhat unusual) food-chain.

New variant CJD

Because of the BSE crisis, a CJD Surveillance Unit was set up in Edinburgh in 1990 to monitor new cases of CJD and detect any increase which might herald a new epidemic. Farmers, veterinary surgeons, and abattoir workers were watched particularly closely because they were most likely to have been exposed to BSE. There has been no increase in CJD in this group. However, alarm bells began to ring in 1995 when two teenagers developed symptoms of brain degeneration which were diagnosed as CJD. These were swiftly followed by four more cases—three people in their 20s and a 30-year-old.[18] Conventional CJD is almost unknown in young people, so was this the start of a new human epidemic due to exposure to BSE, or was CJD occurring in unusually young people for some other reason?

It was possible that the increased awareness of CJD since the BSE epidemic had brought to light a rare form of CJD in the young which had passed unnoticed until the CJD Surveillance Unit was set up. Since there was no test for BSE or CJD at that time, doctors had to rely on case reporting, clinical analysis, and microscopic examination of brain material after death to make the diagnosis. Three criteria convinced doctors that they were seeing a new disease in the UK:

1. CJD in young people was apparently unique to the UK—there were no reports from other European countries or the USA.

2. The clinical course of this disease was slower than classical CJD, with more prominent behavioural changes and memory loss.

3. Perhaps most persuasively, brain abnormalities from these young cases were unique. Experts reported that the spongy cells and deposits of abnormal protein fibrils typical of TSEs, were different from those seen in classical CJD.

This new disease was called 'new variant CJD' and experts released a carefully worded statement: 'in the absence of any credible alternative, the most likely explanation at present is that these cases are linked to exposure to BSE before the Specified Bovine Offal ban was introduced in 1989'.

As the story unfolded, emotive headlines revealed the personal tragedies behind the official figures:

GIVE ME BACK MY LIFE

This is mad cow victim Vicky Rimmer—whose plight could be a fatal time bomb. Vicky, 16—who is blind and unable to talk, move or eat—may be the first known person to have caught the disease from a contaminated burger. 'She was the healthiest child on earth. But she's always loved meat' says mother Beryl.[19]

VEGETARIAN GIRL DYING FROM CJD

Girl who has not eaten meat for more than 4 years has contracted the human form of mad cow disease. The diagnosis of 24-year-old Clare Tomkins's mystery year-long illness has triggered fears that the incubation period could be longer than previously believed.[20]

There is now a test which distinguishes between the abnormal brain protein in BSE and in classical CJD, and new variant CJD cases show the BSE pattern; clear evidence that BSE has infected humans. The ramifications of this are immense and there are many unanswered questions. Can CJD be spread by blood transfusion? Should there be a test available to detect those at risk of CJD? And most particularly, how large will the human epidemic be?

By 1996 there had been a mere 14 cases of new variant CJD diagnosed in the UK and one in France. David Skegg from the University of Otago, Dunedin, New Zealand, in an article entitled 'Epidemic of false alarm' in *Nature*, asked 'if one swallow does not make a summer, do 14 cases constitute an epidemic?'[21] A mathematical prediction of the epidemic, also reported in *Nature*, concluded that:

…until we have a clearer idea of the incubation period the prediction of the number of cases is anything from around one hundred to tens of thousands; that just because only 14 UK cases have been confirmed so far, the epidemic will not necessarily be small, and that even in another four years time the picture will not be entirely clear.[22]

At the beginning of the new millennium there had been just over 100 deaths from new variant CJD in the UK. However, we will still have to wait and see what evolves. In the meantime though it is clear that the BSE epidemic is a truly man-made affair—called 'mad man' rather than 'mad cow' disease by some who point out that in their natural state cows do not make a habit of eating the brains of their dead. There are certainly many serious lessons to be learnt from this unfinished tragedy.

Other man-made diseases

For people in need of new organs—be it a heart, liver, lung, or kidney—the present situation is desperate. There are just not enough spare parts to go round. There are over 5000 people waiting for a kidney in the UK, each being kept alive by a dialysis machine; many will die while waiting for a transplant. The answer to their prayers could be just around the corner. 'Designer' pigs are being specially bred to carry human genes so that the human immune system will not reject their organs, but there is still a major problem—pig viruses. All pigs carry viruses with the potential to infect humans, and although 'clean' pigs can be produced, they still carry endogenous retroviruses. These viruses, which are members of the same virus family as HIV, insert their genetic material into host cell chromosomes and are inherited along with normal DNA. Because of this they would be very difficult, if not impossible, to remove.

In 1997, scientists in the UK found that pig retroviruses released from normal pig cells could infect human cells, showing that in the body these viruses could jump from a transplanted organ to recipient cells, and from there possibly to other humans. In no time we could have a new pandemic on our hands. In the light of these results, several governments put a moratorium on pigs as organ donors until more is known about the potential of their viruses to infect humans. However, in the past pigs have been used to rescue people in desperate situations. Some people for instance have been temporarily attached to pig kidneys instead of dialysis machines; and pig tissues have been used experimentally in humans (for example, islet of Langerhans cells to treat diabetes, and brain cells for Parkinson's disease and Huntington's chorea). When these people were tested for evidence of infection with pig viruses none was found. On this optimistic note, and just when the Council of Europe has called for a worldwide moratorium on pig transplants, many scientists now feel that limited, carefully monitored clinical trials are justified to assess the risk and see the way forward. Otherwise there is a real danger that somewhere in the world poorly monitored trials led by commercial interests could lead to disaster. After all, if this is successful, commercial companies estimate pig organs to be a six-billion-dollar-a-year earner.

Pest control or widespread chaos?

So far all epidemics described in this chapter have been man-made in one way or another, albeit innocently or accidentally. But on occasions we have intentionally released viruses to control pests. The first and most famous use of a virus as a biological control agent was in 1950 when the myxomatosis virus, a natural infection of Brazilian rabbits, was released in Australia and the UK to control the numbers of rabbits (see Chapter 1). This is a marvellous example of selection, both of virus and rabbits. While the virus became less virulent, rabbits became more resistant, so that now myxomatosis in Australian and British rabbits shows the stable pattern of infection seen in their Brazilian cousins.

European rabbits were taken to Australia in 1859 as a source of food and with no natural predators their numbers escalated. They ate native plants and displaced natural wildlife species and now there are an estimated 300 million rabbits creating around 300 million pounds worth of damage to Australian crops every year. With farmers clamouring for action, the Australian Government decided to test out a calicivirus—rabbit haemorrhagic disease virus—which is as lethal and contagious in European rabbits as Ebola is to man. It is spread by many routes including inhalation, so its impact should be far-reaching. This virus suddenly appeared in Europe in the 1980s where it killed over 90 per cent of infected rabbits—so many that now there are fears for rabbit predators like foxes, wild cats, and eagles.

In 1995, rabbit haemorrhagic disease virus was released under strict quarantine conditions in a trial site on uninhabited Wardang island, five kilometres off the coast of South Australia. But somehow, probably as a result of a forest fire driving virus-carrying mosquitoes off the island, it escaped to the mainland. On reaching Flinders Range National Park it killed 750 000 rabbits and very soon it was infecting rabbits all over Australia.

Before release, the virus was tested on 31 Australian wild and domestic animal species and none was susceptible to infection. Likewise, blood tests on workers exposed to the virus on Wardang island were negative. Fears that it will jump species in the wild have so far proved unfounded, so, reassured by this, Australian scientists have released more virus in hundreds of locations around the country. In some areas the rabbit population has already dropped by 95 per cent, but scientists know that they will not be wiped out completely because, as with myxomatosis virus, rabbits will develop resistance.

At the moment, this man-made rabbit infection is doing its job and behaving itself. The Australians have opted for short-term gains when the future cannot be predicted with absolute certainty. There is a heated debate among virologists and ecologists around the world as to whether this is a wise course to take. Some have labelled it 'playing with dynamite' because, as we have learnt from experience, viruses do unexpected things. For this virus to infect other wild animals or man, a mutation would be needed, but even without a mutation, there will be dramatic effects on the local ecology. Already western grey kangaroos, which compete with rabbits for food, have increased sixfold, whereas feral cats, which rely on rabbits as a food source, have declined by as much as 90 per cent in some areas. In addition ecologists are now worrying about birds of prey like eagles, kites, and harriers. Once again we can only wait and see what the future holds. One thing is certain however—this will not be the last of this type of experiment.

Man's very success in the battle for survival on this planet has inevitably disturbed ecosystems which had been delicately balanced for millions of years. This chapter illustrates some of the devastating knock-on effects that these disturbances have caused. Because they are unfamiliar to us, 'new' virus infections are often lethal and highly contagious. Viruses by their very nature will unashamedly exploit all opportunities which arise for their own ends, so we will have to temper our behaviour if we are to avoid new and unpredictable conflicts in the future.

Chapter 3

Coughs and sneezes spread diseases

It is just before Christmas. At her home in Bristol, Mary Smith, a 70-year-old retired shop assistant, is expecting her son John from Hong Kong for the holiday. He arrives feeling tired and feverish, but after a few days of home comforts he is well enough to enjoy a family Christmas with his mother, sister Lucy, and her sons, Richard aged 10, and Jake, two. On Boxing Day, Mary feels unwell and, thinking that perhaps she overdid the food and drink at the family party, retires to bed for a rest. She wakes feeling cold and shivery with aching head and limbs, a dry cough, and a temperature of 38.5°C. Next day she is unable to get out of bed, so John calls the family doctor who says 'It's only a virus. Call me if she's not better in three days.'

The next day Lucy comes to help out, bringing Jake with her. She is shocked when she sees her mother who is by now coughing continuously and having difficulty breathing. She calls the doctor again; pneumonia is diagnosed and Mary is rushed to hospital. Three days later Mary is dead. Probable diagnosis—viral pneumonia.

In London, George Peterson, a 29-year-old member of the ground staff at Heathrow airport is doing a spell of lucrative overtime during the busy Christmas period. He feels ill at work and on his way home he drops in at the casualty department of the local hospital. By the time he gets there he is pale and sweating and has a high temperature. His whole body aches and he is gasping for breath. The registrar, Dr Lewis, is worried by George's condition so he admits him immediately and, since he thinks that George has pneumonia, begins treating him with intravenous antibiotics without waiting for any test results. The next day George is worse; his breathing is laboured and he is showing signs of heart failure. He is transferred to the intensive care unit and is put on a ventilator but he continues to deteriorate and dies in the night. Probable diagnosis—viral pneumonia.

Meanwhile, in the virology laboratories at the two hospitals, staff have isolated influenza virus from nasal swabs taken from both Mary Smith and George Paterson. As a routine, flu viruses are sent to the Central Public Health Laboratory at Colindale, North London for further analysis and here, George and Mary's viruses are found to be identical to each other, but to differ from the type of flu virus currently circulating in the community.

Within a week of these two deaths a dozen more cases of flu are diagnosed, four of them from the same families. Mary Smith's daughter Lucy and her children are taken ill, and within a few days two-year-old Jake is dead. In London, George's girlfriend is in hospital seriously ill, and Dr Lewis and some of George's colleagues at Heathrow are sick. Family doctors in Bristol and West London report a sudden rise in the number of flu cases. Soon the infection is radiating out from these two pockets and spreading like a fog over the whole country, killing many, particularly the young and old, as it goes.

At the Colindale Laboratory, the viruses are typed—influenza HXNY—a new strain only previously identified in the Far East. No one in the UK is immune and the current vaccine does not cover it—we are in for a major epidemic with potentially millions of cases and thousands dead.

This fictitious scenario, although realistic, is fortunately unlikely to happen today because of the global flu surveillance network. But in 1918–19, just as the First World War was coming to an end, this is exactly what happened. The ensuing pandemic killed an estimated 100 million; many more than the Great War itself. Plagues like this represent natural cycles of infection which have had a major impact on human history.

Missionaries and traders

Until Edward Jenner's invention of smallpox vaccination in 1796, natural cycles of infection held sway. Smallpox has now been eliminated as a naturally occurring virus but in its heyday it was a major killer. Because the smallpox virus—as well as measles, mumps, and rubella viruses—only infect man, do not infect the same person twice, and cannot establish a chronic infection, they rely on continual passage from one person to another for their survival. This means in practice that they can only survive where a large population lives in close contact, so that repeated cycles of infection can occur in a constantly renewing pool of susceptible people.

These viruses have probably infected humans since the advent of agriculture led to founding of fixed communities some 10 000 years ago. Around 1000 BC there were flourishing civilizations in the fertile valleys of the Nile in Egypt and the Ganges in India consisting of enough people to sustain the smallpox virus. There are no

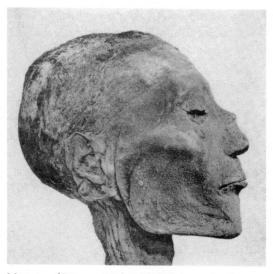

Fig. 3.1 Mummy of Rameses V showing lesions on the face suggestive of smallpox. Reproduced with permission from Dixon, C.W. (1962). *Smallpox* J&A Churchill).

written records of smallpox from that time, but mummified bodies from Egypt, particularly that of King Ramses V who died as a young man, show typical smallpox pustules (Fig. 3.1). The virus probably spread from these strongholds north and east to China, where the first written records of the disease date from AD 340.

From these early beginnings, smallpox travelled with traders, armies, and colonizers to new territories, always taking advantage of virgin populations to infect and causing devastating epidemics wherever it went. It travelled ancient trade routes like the Burma Road and the Silk Route, and was brought to Europe by Islamic invaders and returning Crusaders. By the twelfth century, smallpox was rife in most of Europe, Asia, and North Africa. Here it established the pattern of periodic explosive epidemics which killed one in every three urban dwellers and caused over 400 000 deaths every year in Western Europe.

Between the fifteenth and eighteenth centuries smallpox was carried to the Americas, the whole of Africa, Australia, and New Zealand by early settlers and invaders. In each of these territories the indigenous population suffered terribly because of their total lack of resistance. In some places the disease killed up to half the population, clearly having a significant influence on world history. The defeat of the Aztecs by Spanish invaders is a case in point. Hernando Cortez had under 600 troops when he invaded Mexico in 1520. He was being soundly beaten until he was joined by a small relief force which happened to include someone incubating smallpox. Because this was entirely new to the Aztecs it infected almost all of them and killed a third, so allowing Cortez an easy victory.

Similar stories can be told about other less virulent viruses. Often, like flu and measles, they cause non-life-threatening disease in populations which are familiar with them, but are devastating when introduced for the first time into a society with no previous experience of the disease. Serial epidemics of measles were instrumental in extinguishing the natives of the remote islands of Tierra del Fuego off the southernmost tip of South America where Scottish missionaries, Thomas and Mary Bridges, established an outpost in 1871. A mystery illness, first diagnosed by doctors as typhoid-pneumonia, but soon recognized as measles by Mrs Bridges and the ladies of the mission, arrived on one of the supply ships. The native population suffered severely; as many as half those infected died. European children however only had a mild illness and European adults, who had all had measles in their youth, were unaffected.

In his book, *Uttermost part of the earth,* Thomas and Mary's son, E. Lucas Bridges, gives a first-hand account of the effect that measles had on the Yahgan tribe:

> The natives went down with the fever one after the other. In a few days they were dying at such a rate that it was impossible to dig graves fast enough. In outlying districts the dead were merely put outside the wigwams or, when the other occupants had the strength, carried or dragged to the nearest bushes.[1]

Lucas' father worked day and night burying the dead:

> I remember his going off, Sunday and week-day alike, with pick-axe and shovel on his shoulder, and returning, completely exhausted, late at night. At one settlement, a short distance away from the village, he found a whole family dead, with the exception of one small infant, whom he brought home.

He also describes the devastating after-effects of the epidemic:

> It is astonishing that this childhood ailment, contagious in civilised communities, yet seldom fatal, should wipe out over half the population of a district and leave the survivors so reduced in vitality that another 50% succumbed during the next two years, apparently from the after-affects of the attack.

He goes on to speculate (correctly) on the reasons for the racial differences in the outcome of infection:

> It must be that our ancestors, for generations past, have suffered from periodic epidemics of measles, and we, in consequence, have gained a certain degree of immunity from it. On the other hand, the Yahgans, though incredibly strong, and able to face cold and hardship of every kind and to recover almost miraculously from serious wounds, had never had to face this evil thing, and therefore lacked the stamina to withstand it…It is hardly to be wondered at

that the doctors failed to recognise it for what it was, when it appeared in such a virulent form.

Interestingly he also observed:

> It is worthy to remark that the eight or nine half-breeds in our district, though living exactly the same lives as their Indian kindred, all survived both epidemics and completely recovered their usual health. Doubtless they had inherited from their fathers the power to withstand the devastating fever.

Several such epidemics swept through the islands, and it was as clear to the Bridges family then as it is to us now that measles was a major cause of demise of native tribes in Tierra del Fuego. Lucas Bridges' brother, Will, visited the Ona tribe in 1924 at the time the first 'plague' was brought there by a white family:

> When Will…realised what it was, he advised the Ona to scatter for their lives and hide in the forests as of yore, cutting off all communications with others of their kind. The few who followed this sage advice escaped the first outbreak, only to be caught by the second, which visited the country five years later, in 1929.

Epidemics and pandemics

Epidemics are defined as any unusual rise in the number of cases of an infection in a community and a pandemic is an epidemic which sweeps round the world involving several continents at once. These are the plagues which have been a threat to our existence since ancient times. But understanding exactly why these plagues should regularly sweep through a community and disappear between times is only recent. It dates from a famous measles epidemic in the isolated community living on the Faroe Islands in 1846. Measles had not visited the islands for 65 years, when a carpenter arrived from Denmark eight days after visiting a sick friend. He came down with measles and in the following six months 6000 out of the 7782 islanders also caught the infection. Peter Panum, a young Danish medical health officer, was sent to investigate the epidemic.

By finding that none of the elderly people who had had measles in the previous epidemic of 1781 were sick, Panum established that immunity following natural measles infections is life long. And by carefully noting who his patients came into contact with, he could show that measles is only infectious when the rash first appears, and that the incubation period is 13–14 days. Panum's observations were the first to explain the pattern of childhood infections—that is, why not only measles, but also mumps, chickenpox, and rubella usually occurred in children, and why epidemics appear in a community at regular intervals. Before the immunization programme put a stop to them, these viruses regularly swept through the UK every two years, and although children in isolated rural communities may have been bypassed, most urban children were infected by the age of three.

Influenza

Flu infection is less predictable and more widespread than the common childhood infections. Although flu outbreaks occur every winter, full-blown epidemics only happen every 8–10 years. In a UK epidemic typically up to 10 percent of people get infected and approximately 25 000 die. The last flu epidemic in Britain was in 1989 with 600 000 cases and 26 000 deaths.

Flu pandemics appear at greater intervals and are much more devastating. In the last 100 years there have been three flu pandemics—in 1918, 1957, and 1968—and in each one the virus spread round the world like a Mexican wave. This could only happen because no one was immune—not even those who were infected in previous pandemics.

The term 'influenza', meaning 'influence', was coined by the Italians in the fifteenth century in the belief that flu was caused by a malevolent supernatural influence. One of the earliest documented flu pandemics was in 1562 when it was called 'the newe acquayntance'. Randolph at the court of Mary Queen of Scots in Edinburgh wrote to Sir W. Cecil in London:

> May it please your Honer, immediately upon the Queene's arivall here, she fell acquainted with a new disease that is common in this towne, called here the newe acqayntance, which passed also throughe her whole courte neither sparinge lordes, ladies or damoysells, not so much as ether Frenche or English.[2]

In describing the symptoms he wrote:

> It ys a plague in their heades that have yt, and a sorenes in their stomackes, with a great coughe, that remayneth with some longer, with others shorter tyme, as yt findeth apte bodies for the nature of the disease.

And of fatalities:

> There was no appearance of danger nor manie that die of the disease, excepte some olde folkes.

The reason that we get frequent flu epidemics interspersed with less frequent but more devastating pandemics is the virus' unique ability to change its genetic make-up. The genetic material of flu virus mutates frequently, so that viruses circulating in the community collect two or three changes in their protein structure every year. This is called antigenic drift, because the virus slowly drifts away from its parent stock and, by the same token, away from our immunity built up by previous exposure. Eventually the altered virus is different enough to infect people who are already immune to its parent, and there is another epidemic.

Seen under the electron microscope, flu virus particles come in various shapes and sizes, and all have a fringe of spikes protruding from their surface (Fig. 3.2). These spikes comprise the two proteins known as haemagglutinin (or HA) and neuraminidase (or NA) which are the 'hooks' the virus uses to attach to and get inside host cells. Antibodies against HA and NA block the virus from binding to cells thereby preventing infection and rendering people immune. Thus in molecular terms, antigenic drift is the slow and cumulative minor alterations in the structure of HA and NA which, from time to time, allow the virus to dodge our defences and cause disease. However the high level of partial immunity remaining in the community ensures that antigenic drift will not cause a pandemic.

A flu pandemic occurs when a major change in the genetic make-up of the virus produces an entirely new strain which the majority of the world's population has never met before. This ability for sudden large-scale change, which is unique to influenza A virus, is called antigenic shift and only occurs about every 10–40 years.

Because HA and NA proteins are of major importance in immunity against flu, viruses are named according to the types of HA and NA protein they carry. The type which caused the 1918 'Spanish' flu pandemic was retrospectively called H1N1. It circulated in the community with minor drift until 1957 when H2N2, the so-called 'Asian' flu, appeared. In 1968 this was replaced by H3N2, 'Hong

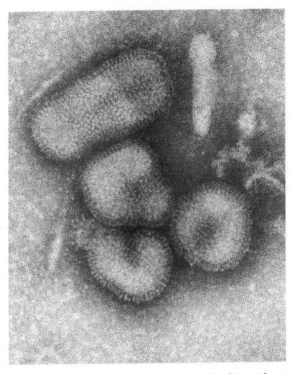

Fig. 3.2 Electronmicrograph of flu viruses showing the fringe of protruding spikes. (Courtesy of Dr Hazel Appleton, Central Public Health Laboratory.)

Kong' flu, and in 1976–7, H1N1 reappeared. Each one of these major shifts coincided with a pandemic.

Research into flu began as far back as 1901—long before viruses were seen or characterized. Eugenio Centanni working in Ferrara, Italy, found that fowl plague (a fatal, flu-like disease of poultry which regularly devastated domestic flocks) was caused by a filterable agent. He carefully mapped the progress of the current epidemic as it moved through Italy, across the Alps into Austria, and thence to Germany. He identified the culprit as an itinerant poultry salesman who was carrying infected stock. This hapless tradesman dropped in at the Brunswick poultry show in the summer of 1901, causing its hasty closure, and dispersing all the participants home with their newly infected stock, to spread the infection further.

Despite these observations, the true identity of the fowl plague virus was not discovered until 1955 when Werner Schafer, working at the Max Planck Institut fur Virusforschung, Tubingen, in Germany, found that it is none other than bird flu virus, a very close relative of human flu. Schafer suggested what later became a proven fact—that under certain conditions flu viruses from different species can undergo a process of gene swapping or genetic reassortment which then enables them to infect another host species and '… in this way a new type of [human] influenza agent might develop from fowl plague and vice versa'.[3]

The genetic material of flu virus is segmented into eight separate genes, so when two different strains of flu virus infect the same cell, the new viruses produced may contain a mixture of the two parents' genes. Many viruses created by this so-called gene reassortment will not find a suitable host to infect or will not be sufficiently different from the parents to cause a problem. But if the reassortment involves the HA or NA genes, then the resulting virus may be different enough from the currently circulating strain to evade host immunity. In that case the new virus will have a selective advantage over its siblings and will spread rapidly with the potential to cause a pandemic.

Genetic reassortment may occur between two human flu viruses infecting the same human cell and this can be made to happen artificially in the laboratory. It can also occur in birds and mammals such as pigs, with reassortment between human and bird or pig flu genes. These animals act as a reservoir of viruses of different strains and from time to time one finds its way into the human population and starts a pandemic.

Piggy in the middle

New strains of flu virus come from the Far East more often than could possibly happen by chance alone. The pandemics in 1957 (Asian) and 1968 (Hong Kong), as well as the Beijing, Shangdong, Wuhan, and Singapore strains all originated there. The favourite theory to explain this geographical puzzle points the finger at rural southern China as the epicentre. Here there are more waterfowl (mainly ducks), pigs, and humans living in close proximity than anywhere else in the world.

For a genetically shifted virus to infect humans, generally a bird–pig–human triangle is needed. Waterfowl are the main natural reservoir of flu viruses because they carry them as harmless infections, and all 15 HA and 9 NA types, in almost any possible combination, can be found in birds. So birds act as the melting pot, throwing up large numbers of different flu strains which are excreted in profusion in their droppings. But only viruses carrying a few of the HA types can infect human cells directly, so for successful spread of reassorted bird viruses to humans an intermediate host is usually required. This is where pigs come in, since pigs are very susceptible to both bird and human strains of flu. The final recombination between bird and human flu viruses occurs in pigs infected with the two strains at once, and what emerges a few times every century is a virus whose genetic material has been reassorted

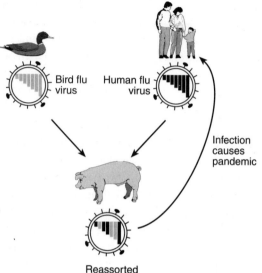

Fig. 3.3 Flu virus reassortment. Bird and human flu viruses undergo reassortment in a pig. This can produce new viruses which infect humans and cause a pandemic.

in a bird, mixed with human flu genes in a pig, and is able to infect and cause a pandemic in humans (Fig. 3.3). The 1957 Asian flu was caused by a human virus with three bird virus genes and Hong Kong flu virus which caused the 1968 pandemic had two.

On the western front?

Everyone agrees that the 1918 flu was exceptional in more ways than one. The pandemic swept through Europe and America reaching the remote wastes of Alaska and the most isolated of island communities.

Half of the world's population was infected and one in twenty of them died. Most fatalities were among teenagers and young adults. This compares with a death rate of one per thousand in other flu pandemics, where most of those killed are either very young or very old.

Although some people died within a day or two of getting ill, as a result of the virus damaging their lungs, most of the deaths occurred about a week later. These later deaths were probably caused by bacterial infection of lung tissue which was already damaged by the virus. Co-infection with bacterium and virus was so common in 1918, when the true cause of flu was still unknown (the virus was not isolated until the 1930s), that the bacterium (which is still called *Haemophilus influenzae*) was considered the possible primary cause of the pandemic. It is now clear that this bacterium is a common inhabitant of the respiratory passages which causes no problems in healthy people. But as soon as cells are damaged by a virus infection like flu, the bacterium can get a foothold, multiply rapidly in the tissues, and cause pneumonia.

The severity of the 1918 pandemic could either have been due to the virus itself or to the unusual circumstances of the time. Europe had been at war for four years. Troops lived in overcrowded, extremely unhygienic conditions, and large groups of them moved rapidly from place to place. These are exactly the conditions which encourage spread of airborne viruses like flu. Many were stressed, exhausted, and poorly nourished—just the people who succumb most easily to infections. But war conditions alone, however terrible, cannot explain the severity of the pandemic, because even in parts of the world untouched by the ravages of war the death toll was unusually high.

Could the virus itself have been particularly virulent? We still do not know the answer to this question, but if similar catastrophes are to be prevented in the future, it is important to find out. Using modern molecular techniques, a group of American scientists recently typed the flu virus extracted from lung tissue of a 1918 Spanish flu victim which had been stored in the laboratory for 80 year, and they came up with some very surprising results.[4] When they compared the 1918 strain with common human, pig, and bird flu viruses they found no similarity to the bird strains, but close identity with one of the pig viruses. To verify this, scientists looked at 1918 flu viruses taken from the body of an Inuit woman who died at Brevig Mission when the pandemic swept through Alaska. Being buried in the permafrost, the virus was as well preserved as if it had been in a deep freeze for 80 years, and this analysis confirmed their earlier findings. Pigs also suffered a flu epidemic in 1918, so the virus could have spread directly from pig to man without any genetic input from birds—if true, a unique observation.

Rather than starting in China, the first cases of 1918 flu were officially reported from near the Western war front in Spain, hence the name 'Spanish' flu. But earlier in 1918 there was an outbreak of milder flu in North America and this may have travelled to Europe with the troops. This raises the possibility that the new strain arose by genetic reassortment between the American and the currently circulating European strain. However, despite the lack of supporting evidence, many virologists still think it likely that the flu virus which caused the 1918 pandemic, like all those which followed, arose in China.

Be prepared

As soon as flu virus was first grown in the laboratory in the early 1940s, the next obvious step was to make a vaccine which would

prevent the disease. Early attempts using an inactive virus preparation were reasonably successful, and vaccination is still the mainstay
of protection against flu today. But there are problems. First, a
vaccine is strain specific, so when a new strain appears the current
vaccine is useless. This was illustrated early on when a new vaccine
was tested on US army recruits in 1945. It gave a good level of protection—only 8 out of 100 recruits exposed to flu got infected. But
in 1947, after a significant antigenic drift, the situation was reversed
and only 9 out of 100 recruits vaccinated with the same vaccine
escaped infection. To some extent this problem can be overcome by
combining different strains in a vaccine. Several strains can be
included without affecting the vigour of the immune response to
each but, of course, to be effective the cocktail has to include the
coming winter's strain and this requires a certain amount of guess
work.

The second problem is that the vaccine is difficult and expensive
to produce. Even today the virus must be laboriously grown in
chicken's eggs, purified, and subjected to a battery of efficacy tests
and safety checks before it can be released for use. So production
takes several months and each vaccine contains only three strains.

With these problems in mind, the World Health Organisation
(WHO) set up the Global Initiative in 1947. This is like a worldwide
spy network involving 110 laboratories in 85 countries, but the difference between this and real espionage is that there is co-operation
between all the countries involved and mutual benefit. Each has a
National Influenza Monitoring Laboratory, and modern flu prevention still relies on their surveillance activities. Each National
Laboratory is at the heart of an information network whose tenta-
cles radiate out into the countryside, feeding information back to
the centre. They constantly monitor the flu strains circulating in

their particular region, identify new strains, and regularly report back to one of the three International Laboratories which are situated in London, Atlanta, and Melbourne. In Britain, up-to-date information on flu viruses flows from the community to the laboratory via various sources: 100 designated general practitioner 'spotter' practices report all flu-like illnesses; school medical officers record illnesses in school children; the hospital emergency bed service sends information about hospital admissions; and the Office of Population, Censuses, and Surveys provides death registration information. All this data is collated by the Public Health Laboratory Service Influenza Surveillance Team at Colindale in London and published once a week.

The WHO is now investing heavily in China in the hope of identifying new potential epidemic or pandemic strains of flu as they arise, so acting as an early warning system for the rest of the world. Ten special laboratories have been set up which are dedicated to flu virus isolation in China with rudimentary sentinel physician networks to feed in clinical information. At the research level, farmers, abattoir workers, and their animals are being monitored as a likely source of an emerging new strain.

In February each year, scientists from the three international WHO laboratories meet in Geneva to consider their data and recommend the cocktail of flu strains to be included in the worldwide vaccine for use the following winter. Once this is agreed it then takes 6–8 months to prepare the vaccine. In October each year new vaccine is dispatched to doctors in the Northern hemisphere and the vaccination campaign begins. Somewhere in the region of six and a half million doses are distributed in the UK, which is enough to cover those at most risk from severe flu including the elderly and people with chronic diseases, particularly heart or lung diseases, diabetes, and kidney failure. This practice cuts the death rate from flu by half.

It is astonishing and at the same time comforting to think of this massive surveillance programme continuously going on behind the scenes. But will it really prevent the next pandemic? Dr Maria Zambon, who heads the Flu Laboratory at the Central Public Health Laboratory, thinks it will.[5] She cites the emergence of 'Wuhan' flu as an example. A number of Wuhan (H3N2) viruses found circulating in China in September 1995 were sent to Atlanta and London for detailed typing and to test the level of protection afforded by immunity to the currently circulating 'Johannesburg' strain (also H3N2). By the time 'Wuhan' reached Hong Kong in October 1995 and Singapore a month later, scientists knew that it had drifted to such an extent that its protein sequence differed from 'Johannesburg' proteins by 4 per cent. At this level most people immune to 'Johannesburg' would have some protection against 'Wuhan'—it would not cause a pandemic but an epidemic was on the cards.

During 1996, 'Wuhan' reached the East and West coast of the USA, and appeared in Europe via Norway. In February of that year 'Wuhan' was recommended for inclusion in the vaccine which was ready for distribution in September. In November, the 'Wuhan' virus was found in the UK. Although on this occasion there was no epidemic, it serves to illustrate how little time the UK has to get prepared. In theory, the further a country is from China, the more time it has in hand but, as we see in the story at the beginning of this chapter, these days a new flu strain could reach any country in the world in an airborne human incubator in less than 24 hours.

A pandemic by 2010?

Flu pandemics occur on average every 10–40 years and since the last one was in 1968 we are surely in for another one soon.

In May 1997, a three-year-old boy was admitted to hospital in Hong Kong with an acute respiratory illness and a high fever. A few days later he was dead, but not before the offending flu virus was isolated from his respiratory tract. This episode would have passed unnoticed but for the fact that scientists in the Hong Kong Department of Health could not identify the strain of flu virus. They sent it to all three WHO International Reference Laboratories and they all came up with the same answer, H5N1—an entirely new strain of virus to infect humans, but identical to the one which killed 4500 chickens on three Hong Kong farms earlier in the year. At that time, H5 viruses have never infected man. In fact, no bird flu virus had ever been recorded as infecting man directly, without the intermediate pig 'mixer'. The WHO sent a team of experts to Hong Kong to help with the investigation and then everything went quiet. That is, until December when suddenly hundreds of chickens in a Hong Kong market died. H5N1 was again the culprit and more infected chickens were found on a farm near the Chinese border. By this time there had been 17 human causes of H5N1 infection and four more deaths.

'World facing killer bug'

the newspaper headlines screamed. With panic mounting among its local inhabitants, and a rapidly falling tourist count, the Hong Kong Government decided to act. All Hong Kong's 1.2 million chickens from 200 farms and 1000 trading posts were slaughtered within 24 hours. Imports from China were banned, which left Hong Kong devoid of fresh chickens for its approaching traditional New Year banquets and, presumably, full of irate chicken farmers, traders, and cooks.

Did this prompt action prevent a crisis? We do not know, but at the very least it prevented the virus infecting more Hong Kong chickens where it would have the chance to reassort with other chicken viruses. Although scientists have now analysed its genes, there is still a lot we do not know about this new flu strain. Why is it so lethal in chickens (it killed 23 out of 24 experimentally infected birds)? How does it infect humans? Can it be passed from one person to another, or did all the human infections in Hong Kong come directly from chickens? This time an epidemic or pandemic has been averted, but we cannot afford to relax our surveillance.

Poliomyelitis

A young man with drop foot, beautifully depicted on a funerary stele dating from the eighteenth Egyptian dynasty (1580–1350 BC), is perhaps the earliest record of a virus infection and shows us that paralytic polio is an ancient affliction (Fig. 3.4). However, the disease only became a significant public health problem in the twentieth century and then mainly in industrialized countries. In Europe and North America, epidemics reached a peak in the 1940s and 1950s and then abruptly disappeared just as polio began to emerge as a major health problem in developing nations. Although the disease is now being successfully eradicated worldwide, hot spots still remain in India and Pakistan. These remarkable changes in disease pattern in less than a hundred years were caused by changes in man's lifestyle which profoundly affected the biology of the virus.

Polio virus, like flu, has to pass rapidly and continuously from one host to another in order to survive. But, devastating as it was to the people affected, polio was always a relatively rare disease and never caused epidemics or pandemics on the scale of flu.

Fig. 3.4 Boy with drop foot, probably caused by polio, from an Egyptian tomb of the eighteenth dynasty.

Before vaccination was introduced in the 1950s, polio was a much-feared disease. It usually struck during the summer months, mostly affecting children of affluent families. The illness often started without warning, giving headache, fever, vomiting, and a stiff neck as the virus spread through the body. Then it targeted cells of the central nervous system, particularly the spinal cord, damaging the nerves to such an extent that a child who was perfectly healthy one week could be permanently paralysed the next. Anything from part of a single muscle to virtually the whole body could be affected. When the virus attacked the nerve supply to the muscles used for breathing, victims were condemned to life inside an iron lung. Many well-known people suffered from polio, among them Sir Walter Scott who recovered but for a limp, and Franklin Roosevelt who, from the age of 40, was permanently paralysed below the chest.

The history of polio research is an unhappy one, and this goes a long way to explaining why it took nearly 50 years from the demonstration of a virus to the production of a vaccine. In 1908 Karl Landsteiner and Edwin Popper, working in Vienna, found the virus by using a filtered extract from the spinal cord of a nine-year-old boy who died of the disease to transmit polio to two rhesus monkeys. Thereafter, polio infection of rhesus monkeys was used as a model for the disease in humans in virtually all polio research until 1948. Then, John Enders and co-workers at the Boston Children's Hospital, USA, finally succeeded in growing the virus in tissue culture—a feat which won them the Nobel prize in 1954.

As rhesus monkeys were so expensive, very few institutes could support polio research, and for 25 years the field was dominated by a group at The Rockefeller Institute for Medical Research, USA, headed by Simon Flexner. Monkeys could only be infected through the nose, and Flexner's research was biased by his firm belief that this was also the natural route of infection in humans. Because he thought that polio was primarily a disease of the central nervous system, he assumed that the virus, entering the body through the nose, thence made its way straight to the brain. He was so sure of this that during the polio epidemic of 1937 in the USA, many people were persuaded to have their nasal passages lined with zinc sulphate paste to block infection—uncomfortable, unsightly, and, alas, without any beneficial effect.

Carl Kling was 24 and working as an assistant in the Bacteriological Institute in Stockholm in 1911, when there was a polio epidemic in Sweden. He detected the virus not only in the noses of polio patients, but also in their mouths and the wall and contents of their guts. More surprisingly, he also found the virus in the guts of healthy family members of patients and even in healthy members of the public. He reported these discoveries in 1912, but rather than revolutionizing contemporary thinking on polio as primarily an infection of nerve cells, his findings were assumed wrong and totally ignored. Until that is, some 26 years later, in 1938, when John Paul, a physician at Yale University in the USA found the same thing. So, for a quarter of a century the field of polio research was misled because of the influence wielded by the one group able to afford the only—and, as it turned out, misleading—animal model available. When this 'new' information was taken on board all the pieces of the polio puzzle began to fall into place.

We now know that polio virus spreads by the faecal–oral route. It enters the body through the mouth and, although it does not cause gastroenteritis, the virus grows in the cells lining the small intestine. Large amounts of virus are excreted in the faeces of infected people, whether they have symptoms or not. Infection is not a rare event, as previously thought, but common, though usually symptomless. Paralysis is an exceptional manifestation of this common infection. Doctors have calculated that only somewhere between one in a hundred and one in a thousand infections result in paralysis, although the reason for this is not understood. With this knowledge, the patterns of infection immediately became clear.

Polio virus can survive for several weeks in water and sewage, and thrives in overcrowded conditions with low standards of hygiene. Consequently, until the first half of the twentieth century, polio virus was endemic in developing countries and in poorer areas of industrialized countries. Here, infection occurred very early in life, and by the age of five years virtually all children were immune. In this situation the paralytic form of the disease was rare, but in the more affluent societies of Europe and the USA, where standards of hygiene were high, young children were often protected from infection. Under these conditions a large enough susceptible population periodically built up, epidemic conditions prevailed, and the virus spread through the community.

But still polio epidemics differed from flu and measles in being more geographically confined and affecting relatively small numbers of people. The outbreak of polio in the summer of 1954 in an affluent community in New Canaan, Connecticut, USA, for example, centred on a nursery school, with all infections occurring in the school children, their parents, families, and friends. Antibody studies showed that virtually everyone in the community was infected by the virus that summer, but only 16 cases of paralytic polio occurred. This pattern of infection is aptly referred to as 'the iceberg phenomenon' because most infections are below the level of detection; just those in the tip get paralytic disease. Geographically, the infection did not spread beyond the small community centred on the nursery school because the high standards of sanitation and hygiene denied the virus its usual route of spread. On another occasion, in 1955, 120 000 children in the USA were given polio vaccine which had not been adequately inactivated and was therefore still infectious. Of the 50 per cent who did not already have antibodies to polio virus, only around 1000 became infected and 600 developed paralytic polio.

In developing countries, the change from endemic to epidemic pattern of infection seen during the second half of the twentieth century was the inevitable consequence of rising standards of living. Infection in early childhood was prevented, thereby exposing a susceptible population to infection later in life and giving rise to epidemics of paralytic polio which had not been seen before.

Introduction of the vaccine in 1955 had a dramatic effect on epidemic polio in the industrialized world. In the USA, the incidence of paralytic polio immediately dropped tenfold, from the pre-vaccine level of around 20 000 to 2000–3000 cases per year. Similar falls followed in many other countries and in 1988 the WHO global eradication programme made vaccine available worldwide with the declared aim of polio eradication by the year 2000. In 1991, Greenland, Western Europe, and Australasia were officially declared polio-free, followed by the Americas in 1994, and China in 1995. Although pockets of infection still persist in India, Pakistan, Russia, Africa, Indonesia, and the Middle and Far East, hopefully, within the next decade polio, like smallpox, will only be a receding memory.

The common cold

Throughout the world, colds strike more frequently than any other virus infection and in the majority of cases rhinovirus is responsible. Most people get the familiar symptoms—runny nose, swollen watery eyes, and general stuffed-up feeling—about twice a year. So rhinovirus infection, although apparently trivial, is of major economic significance, being one of the commonest causes of absence from work. But despite this, and decades of research on the virus, there is still nothing to prevent or cure it.

In 1930 Alphonse Dochez and colleagues in New York managed to transmit colds from humans to chimpanzees using filtered nasal washings from a person in the early stages of a cold. The infection then spread among their colony of eight young chimps. Following this they transmitted the infection to human volunteers and their hopes of isolating a virus were high. But their early claims to have grown it in culture could not be repeated and no further progress was made until Sir Christopher Andrewes, in London, took up the challenge. In 1932 he started in a small way, at St Bartholomew's Hospital, using medical students as volunteers (because, he said, they were cheaper and easier to get hold of than chimpanzees). In 1946 however he decided to bid for a complete isolation unit. In the bleak surroundings of Salisbury Plain, Wiltshire, he found what he was looking for—a hospital built by the Americans in 1941. Originally intended to house the victims of the epidemics they confidently expected as a result of war-time bombing, this was later used as an army hospital when no epidemics occurred. With the war over, the Americans had no further use for the building and they generously handed it over.

Andrewes opened the Common Cold Research Unit in late 1946 and for several decades generations of volunteers spent 10 days in fairly basic accommodation, with only a little pocket money and a 40 per cent chance of catching a cold. Amazingly they had no shortage of applicants—in all, 11 000 human guinea pigs passed through, including honeymoon couples and some who apparently enjoyed the experience so much that they came back again and again. Fifteen visits was the record.

Inmates were housed in pairs and the rules were strict. Arrivals were on Monday and the quarantine period (necessary to detect a cold already contracted) lasted until the following Saturday. Tuesday included the excitements of a chest X-ray, a medical examination, and a pep talk. Then life settled into a daily routine of visits from matron and the doctor and recording symptoms. Inmates were allowed out of their rooms but had to observe the '30-foot rule'—that is, they were not allowed to approach within 30 feet of another human being. In his book entitled *In pursuit of the common cold*, Andrewes relates the dire consequences of breaking the rules:

> On very infrequent occasions volunteers have flagrantly and repeatedly ignored the rules, as when members of opposite sexes have associated together: and then it has been necessary to dispatch them home forthwith, withhold their pocket money and exclude them from the trial.[6]

If Saturday arrived without a cold then the lucky volunteers got to lie down with head tilted back while a gowned and masked figure dropped unidentified fluid up their noses. Thereafter the daily inspection of paper hankies became an important ritual. Once used, these soggy items had to be stored as vital evidence of a cold. Final release was on Thursday and home on Friday.

In the laboratories attached to the Unit, desperate attempts were being made to isolate the culprit virus. Somehow it had to be grown outside the human body and since other viruses could be grown in eggs, this seemed like the best place to start. When this technique failed they tried others. After the successful growth of polio virus in human embryo cells in 1948, many different embryonic tissues were tested. When these were in short supply in the UK they were flown in from Sweden (where therapeutic abortion was more commonly practised). Despite all this effort, again no luck.

With the lack of success in culturing the virus, attention turned to trying to grow it in animals. Ferrets, which could be infected with flu virus, as well as the usual small laboratory animals used at the time—mice, rats, hamsters, guinea pigs, rabbits—were tried, but with no success. Whenever Sir Christopher heard reports of an animal with a cold he followed it up. Attempts were made to infect squirrels (both grey and flying varieties), voles, hedgehogs, kittens, pigs (later eaten by the staff—a welcome relief from post-war rationing), and various species of monkey and baboon—but all proved resistant. The capuchin monkeys (obtained especially from South America) seemed to develop watery eyes after infection—a promising sign until they realized that this is a constant feature of the animals, alternatively called 'weeping monkeys'.

Early attempts at the Unit to investigate the infectiveness and spread of the virus were also dogged with problems. Cross infection between two inmates of the same room was very rare and Sir Christopher concluded that their volunteers were too resistant to the virus to be useful in these experiments. What was needed was an isolated community that had not met the virus recently. Consequently, in the summer of 1950, they headed off to the uninhabited Island of Eilean nan Roan (Isle of Seals) off the coast of Sutherland in the north of Scotland. This island had been abandoned by its inhabitants in 1938, but the houses were renovated to accommodate 12 volunteers—10 students and an ex-policeman and his wife who were persuaded to go along to look after the party.

The group had spent 10 uninterrupted weeks on the island when, in mid-September, they divided into three groups ready for a visit from six people freshly infected with colds at the Salisbury Unit. Various strategies were acted out in an attempt to find out how the virus spreads. For several hours the six cold sufferers occupied an empty room which was then occupied by the first group of islanders. Next, the six spent time in part of a room, separated by a hanging blanket from the second group of islanders. Finally, the six with colds and the third group of islanders lived and ate together. The results were at the very least disappointing. To quote

Sir Christopher 'To our astonishment and dismay not a single cold developed in our islanders despite close contact'. But he also adds that the whole venture only cost £1500, and that 'there were more than 12 people who derived immense enjoyment from a venture planned in the interests of science'.[6]

During the 1950s many new viruses were isolated and several of these were tested at the Salisbury Unit to see if they caused colds. Among them were some which gave similar symptoms but differed slightly from the classic cold. Adenovirus, discovered in 1953, has 47 subtypes which infect man, some of which give cold-like symptoms but with a slightly longer incubation and a more severe illness. Similarly echo, Coxsackie, and parainfluenza viruses, all of which have multiple subtypes, could produce cold-like symptoms in volunteers.

It was not until David Tyrrell joined the Unit in 1957 that the breakthrough came. He made some minor but critical changes to the virus culturing conditions, which included lowering the temperature of the incubator from body temperature (37°C) to 33°C (the temperature of the nasal passages where the virus grows naturally). This led, at last, to the successful growth and subculture of a virus which gave the symptoms of a cold in volunteers. In 1963 this virus was officially named rhinovirus from the Greek word 'rhis' meaning nose.

We now know that not only are there 100–150 different types of rhinovirus, but that the common cold can also be caused by several other viruses each of which have multiple strains or subtypes, so that there are in reality several hundred 'cold' viruses. This explains the early conflicting results at the Salisbury Unit since they were probably working with many of these different viruses. It also explains why there is no vaccine for rhinovirus and why we are visited by it with monotonous regularity.

The acute virus infections described in this chapter amply demonstrate the enormous advances we have made in the fight against viruses in the last few decades. Vaccination has assured that we no longer see the familiar cyclical epidemics of many acute infections which killed and maimed in the past and children world-wide have benefited. But still with viruses we have to expect the unexpected. We are due for a flu epidemic soon but no one can tell when it will come, whether our monitoring systems will detect it in time, what virus strain will be involved, or whether the current vaccine will protect against it. Once again we are at the mercy of the virus.

Unfortunately, not all viruses can be defeated by vaccines and in the next chapter we shall explore some of the more intransigent chronic and persistent virus infections.

Chapter 4

Unlike love, herpes is forever

Anna Grant is a working mum. From 10 a.m. to 3 p.m. each day she is a secretary; the rest of the time she is a housewife and mother. During term time she takes her three young children to school on her way to work and collects them on her way back. In the school holidays her mother looks after the children. These arrangements work very well. Anna's mother, Pat Allan, is a healthy, energetic 60-year-old widow who loves seeing her grandchildren and looks forward to the school holidays when they spend most of the day with her.

Now it is the summer holidays. One afternoon, as Anna collects the children, she notices a small red area high on her mother's left cheek and asks what it is. Pat doesn't know, but she says that she has noticed tingling in the left side of her face for the last few days. However, feeling perfectly well, Pat is ignoring it. She kisses the children goodbye as usual and then gets ready to go out and visit friends.

When she wakes up the next day Pat is surprised to see that there are now several red patches on the upper part of her left cheek. On closer inspection in the mirror she can see that they are patches of tiny blisters. The tingling is still there and her left eye has now begun to sting but she still feels well so she is not particularly concerned. The children arrive in the morning as usual, and it is not until Anna questions her in the afternoon that Pat admits that the rash is spreading and is now accompanied by shooting pains. Her left eye is particularly painful and is beginning to swell. Anna is alarmed and persuades her mother to go to her doctor. The doctor takes one look and diagnoses shingles. He says there are very good tablets to control the rash and gives her a prescription for acyclovir and some painkillers. He warns her that this may take some time to clear up and suggests that she takes it easy for a while. Since she has conjunctivitis in her left eye, he sends her to the ophthalmology department at the local hospital to have it examined. Luckily, although painful, there are no complications and the rash gradually disappears.

Anna and her children are about to go on their summer vacation so she makes alternative arrangements for the children for the remainder of the week and then they set off for the seaside. The children join a holiday camp full of kids of their own ages. They have a glorious two days, but then things start to go wrong. On the third morning all three of Anna's children wake up cross and listless. They don't want to do anything and it is all Anna can do to persuade them to go to the camp. Eventually they go, but they return at lunchtime complaining of headaches, sore throats, and watery eyes. By this stage they don't look well so she takes their temperatures and finds them to be high. Anna feels perfectly well herself so suspects that they have caught colds from the kids at the camp. But the next day she is proved wrong—they are covered from head to toe in red spots. She calls the doctor—chickenpox is diagnosed and their holiday is at an end. Two weeks later most of the other kids in the camp have chickenpox. How did this happen?

Both chickenpox (also called varicella) and shingles (also called herpes zoster) are caused by the same herpes virus—varicella zoster virus (VZV). When VZV infects for the first time it causes chickenpox. You are then immune and will not get chickenpox again, but the virus stays hidden in your nerve cells for life. Occasionally, often several decades after the original chickenpox, the virus pops out of a nerve cell, and infects the area of skin which that nerve supplies. This causes the localized rash of shingles—most commonly on the face, as in Pat's case. Here, the virus is hidden in cells of the trigeminal nerve, and if it is in the ophthalmic branch of that nerve then the rash involves the eye. Nerves supplying the skin of the chest and abdomen emerge from each side of the spinal cord and run round the left and right sides of the body to meet and shake hands with each other in the centre front—like a bear hug from behind. So shingles on the trunk forms a band around one or other side of the body, stopping abruptly in the middle. The words 'shingles' and 'zoster' mean 'a belt', and 'herpes' is derived from the Greek word meaning 'to creep like a snake'—presumably all describing the distribution of the rash. The Norwegians are even more explicit; their word for shingles refers to the pain and literally translates as 'a belt of roses from hell'.

The tiny blisters of shingles, just like the spots of chickenpox, are packed with virus so people with either form of the disease are highly infectious to those who have not had chickenpox. However, no one catches shingles from chickenpox because the virus that causes shingles comes from inside rather than outside the body.

Viruses like VZV which hide in your body for life are known as persistent or latent viruses. These have quite different habits from the acute viruses described in Chapter 3 which live nomadic lives and are constantly on the move in search of a temporary, susceptible host. To have the best chance of being transmitted to another host, acute viruses have to produce as many new viruses as possible during the brief 5- to 10-day interval between initial infection and elimination by the immune response. Persistent or latent viruses have quite a different approach to life. They put their efforts into hiding from the body's immune system so that they can spend long periods of time colonizing a single person. And since they do not have to rely on the rapid passage from one susceptible host to another for their existence, they can survive where the population is sparse and contacts between people are few. Many were endemic in remote and isolated communities, like the Indians of the South American rain forests and the highland tribesmen of Papua New Guinea, before these people made contact with the outside world.

Many persistent viruses are much more ancient than the human race. Humans inherited these from their primate ancestors where co-evolution of virus and host had led to mutual understanding which now only rarely becomes unbalanced and leads to significant disease. Hence one of the features of persistent viruses which contributes to their success is that under normal circumstances infections are not life-threatening. It would be counterproductive for these viruses to kill off their long-term host, so they tend to cause a mild illness, or even no disease at all, when they first infect, slipping into the host and taking up residence virtually unnoticed.

The sleeping beauty

Viruses like herpes simplex (HSV) and VZV have two separate life cycles. In one, called a latent infection, they are completely hidden inside a cell. In the other, they reproduce and spread—this is called a productive infection. They manage this double life by infecting two different types of tissues—skin and nerves. In skin cells, HSV and VZV can grow and reproduce, whereas in nerve cells, they generally cannot. They spend most of their time in nerve cells completely invisible to immune attack, in no danger of being sought out and eliminated.

There are two types of HSV—1 and 2. Both are spread by close contact; HSV-1 by kissing, causing a cold sore on the face; HSV-2 by sexual contact, causing genital herpes. HSV-1 first infects through a tiny, unnoticed break in the skin surface—usually near the lips. It then grows locally in skin cells causing the familiar sore. The virus carries on infecting more skin cells until the immune response puts a stop to it. But during this growth phase the virus also infects local nerve fibres. Usually, when this virus infects a cell its genetic material makes straight for the nucleus, a distance of a few microns. But in a nerve cell the nucleus is located in the grey matter of the brain or spinal cord, so the virus has a long way to go. It travels along the nerve fibre towards the central nervous system at a rate of about 10 millimetres per hour, and when it finally gets to the nucleus it finds that it cannot go through its normal virus-making cycle because the chemicals it requires for the process are not there. But this does not mean that it was a wasted journey. Like a seed lying dormant in the soil, waiting for the right conditions to bring it to life, the virus lodges there in an inert form.

Meanwhile, back at the skin, the infection attracts the attention of immune T cells which gather to eliminate it. This they can do without much trouble, and the sore heals up and is forgotten. Antibodies made in response to the infection prevent reinfection with HSV-1 in the future, but neither immune T cells nor antibodies

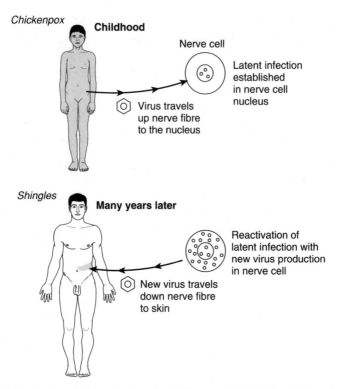

Fig. 4.1 Latency and reactivation of varicella zoster virus. During chickenpox, the virus travels from infected skin up nerve fibres to the nerve cell nucleus in the dorsal root ganglion where it remains in a latent form for life. In reactivation, the virus travels down the nerve again to cause the rash of shingles.

can rid the body of the virus hiding in nerve cells. Since nerve cells cannot provide the necessary chemicals to kick-start virus reproduction, the life cycle of the infection goes on hold and no viral proteins are made to alert the immune system of its presence. So the virus can just sit there completely undetected for the lifetime of the host. VZV uses exactly the same strategy as HSV to establish a latent infection, but because the rash of chickenpox is more widespread than a cold sore, more nerve cells carry latent VZV (Fig. 4.1).

For this latent type of infection to be successful, the virus must hide in a very long-lived cell. Nerve cells are ideal because they do not divide and do not age or die—those we are born with have to suffice for the whole of our lives. So there is little chance of the virus getting lost, as it might if it hid in a rapidly dividing cell with a short life span.

But to survive in the long term, the virus must complete its life cycle and produce more viruses which can get out and infect a new host. If not, the virus will die along with its host.

The reawakening

Latent viruses have periodic bursts of activity called reactivations, when thousands of new viruses are produced. The months or years of silence between these episodes maximize the chances of finding new and susceptible people to infect—a recent friend, partner, or perhaps a new baby in the family. This strategy is so successful that HSV and VZV lodge in almost everyone. Just under half of all adults regularly suffer from cold sores and most people have shingles at least once during their lives. The fictitious story of VZV reactivation and spread given at the beginning of this chapter is actually quite a common scenario. Similarly, for HSV, the classic route of spread is from an adult with a cold sore kissing a young child who may then spread the infection to other children by sucking or dribbling onto shared toys.

The way in which HSV and VZV reactivate from latency is not at all clear, although certain triggers are common. Many people know what reactivates their own latent HSV—be it fever, stress, a cold, or menstruation. Ultraviolet light is also a frequent trigger, and some unfortunate people get a cold sore every time they go on a skiing holiday. Whatever the trigger, somehow the virus is suddenly able to grow in the nerve cell in which it has been dormant for so long. New viruses travel down the nerve fibre, infect the skin, and a new cold sore (HSV-1) or shingles rash (VZV) appears. The same process is repeated at each reactivation and with HSV, as the same nerves are involved each time, cold sores in one person always appear in the same place.

Despite the fact that latent infections are thought to be entirely invisible to the immune system, immune control must play some part in preventing reactivation because people with impaired immunity, particularly those with T cell defects, often have major problems with HSV and VZV. In AIDS patients, for example, reactivated HSV can cause a generalized skin infection and may even invade the brain causing fatal encephalitis. Shingles can appear at any age, but usually it occurs in old age when immunity is waning. AIDS and cancer patients often suffer repeated severe attacks of shingles, sometimes involving more than one nerve at the same time. The rash is most common on the face where, if it includes the eye, it can cause blindness.

Hide and seek

Unlike HSV and VZV, most persistent viruses do not have entirely separate latent and productive life cycles. Both phases can occur in the same cell but under different conditions. In this case there are probably always cells somewhere in the body producing virus, so latency is never complete. Their presence attracts notice and so these viruses constantly have to dodge host immunity.

Two common viruses—cytomegalovirus (CMV) and human papillomavirus (HPV)—both choose as their long-term home cells which are actively dividing but which never age or die. These are called stem cells, and for the lifetime of the host, they constantly replenish those cells which are lost through ageing or wear and tear. The DNA of CMV and HPV beds down in the nucleus of these cells and reproduces along with the cell's DNA. While one copy of the virus' genetic material remains in the stem cell, the other transfers to the 'daughter' cell which is eventually destined to die.

HPV causes warts by infecting the stem cells of the innermost layer of the skin. From here, maturing skin cells work their way up to the surface where they die and form part of the insensitive outer surface which is eventually shed. HPV is latent in stem cells, but as the daughter cells mature, their internal composition changes and they are able to provide the necessary chemicals for HPV to reactivate and reproduce. So by the time a mature infected cell reaches the outer surface of the skin and is about to be shed, it contains thousands of viruses ready and waiting to spread to other people.

CMV generally infects early in life but causes no disease. This virus seeks out those bone marrow stem cells which mature into macrophages—a type of blood cell which patrols the tissues looking for and removing any foreign or dead material. So CMV is transported all over the body and in certain tissue sites, particularly the lungs, the virus can reactivate. In people with normal immune systems, viral proteins on the surface of infected cells signal to the immune system and are quickly recognized as foreign. The cell is then killed by immune T cells before it has had time to make new viruses, so no disease ensues. But in people with immune deficiency, CMV can spread into the tissues and cause severe, even fatal disease.

CMV infection of the lungs—pneumonitis—is a constant problem in people whose immune system is suppressed. This is particularly so in recipients of bone marrow or organ transplants, because they have to take drugs to suppress their immune system and prevent their organ being rejected. CMV pneumonitis is especially common in CMV-negative people who get a transplant from a CMV-infected donor because the virus is very likely to be transferred along with the donor organ. Then, CMV infection occurs for the first time immediately after the transplant, when the dose of immunosuppressive drugs is high, causing death in 80–90 per cent of untreated cases. CMV infection during pregnancy can also cause problems. The virus may cross the placenta and infect the developing foetus. If this happens early in pregnancy then it can cause severe abnormalities of the developing brain, lungs, and liver—but fortunately this is very rare.

For persistent viruses, evading immune recognition is the key to survival, and they show astonishing ingenuity in establishing and maintaining a hold on their host. Even if they go into hiding inside a cell and are completely indistinguishable from the host, from time to time they have to reactivate to ensure their long-term survival by massive production of virus. So these viruses have evolved a variety of mechanisms designed to extend their reproductive phase for as long as possible by subverting or suppressing the immune response directed against them. Since T cells are the most effective weapon which the body has against viruses, it is here that they concentrate their efforts.

All mammals have T cells which are capable of killing their own cells, so these have to be carefully controlled and kept in check. One way is to ensure that T cells can distinguish between self proteins and foreign proteins. In each cell a complex series of reactions breaks proteins up into short pieces called peptides and displays them on the cell surface. If these are recognized by T cells as self peptides then the cell is left intact. But in a virus-infected cell, viral proteins enter this pathway and are broken up into peptides. These viral peptides displayed on the cell surface act as flags to alert T cells to the fact that they are foreign and to kill the cell. So if viruses could prevent this display of their peptides they could perhaps camouflage their presence.

The pathway which converts a protein in the cell to a peptide on the cell surface involves many different molecules with individual functions and viruses may strike at any key point to interrupt the whole process. Interestingly, several completely unrelated viruses have evolved very similar mechanisms for subverting these pathways. First of all the protein must be chopped up into peptides by enzymes which act like molecular chainsaws, and a herpes virus known as Epstein–Barr virus (EBV) manages to avoid this very effectively. The virus needs just one protein to help it remain latent inside a host cell. This is an unusual protein which the cell's enzymes cannot break down into peptides of the size that T cells will recognize, so cells in the body carrying EBV which express only this protein go unnoticed.

The next step in the process involves special transporter molecules (called TAP) which escort the peptides to the endoplasmic reticulum (a compartment which specializes in assembling molecules for dispatch to the cell surface). Here each peptide is attached to a carrier molecule which accompanies it to the cell surface and holds it there ready for recognition by, and then a lethal hit from, a passing T cell. Herpes simplex virus tricks the cell by making a protein which binds to TAP and inactivates it. This prevents herpes simplex-derived peptides entering the endoplasmic reticulum and being available for transport to the cell surface; hence new viruses can be made in the cell without immune recognition by T cells.

CMV makes two proteins which somehow induce the cell to use its enzymes to chop up its own molecules which would normally carry viral peptides from the endoplasmic reticulum to the cell surface. It is not entirely clear how this process works, but in a CMV-infected cell, newly made carrier proteins are degraded in less than a minute, and so very few CMV peptides reach the cell surface.

These are just a few viral survival strategies out of the many which have been uncovered recently; assuredly there are many more which we still know nothing about. However, although each of these viruses appears to have an infallible counter-attack to the onslaught of the immune system, co-evolution has ensured that most persistent viruses establish a balance between successful virus production and immune recognition. The balance may be set slightly differently for each virus, so that some viruses are easier to find in the body than others, but usually the virus infection survives at a level which does not inconvenience the host. The exception to this is human immunodeficiency virus (HIV) which, probably because it has only recently infected humans, has not yet established this life-long balance but instead kills most that it infects.

A reverse turn

Retroviruses like HIV are so called because of their unique survival strategy. Their genetic material is carried as RNA, whereas in all organisms except RNA viruses the genetic code is carried as DNA which is only translated into RNA as a prelude to making protein from an individual gene. Along with their RNA, retrovirus particles carry an enzyme called reverse transcriptase (RT). Once inside a cell, this unique enzyme can reverse the normal flow of DNA → RNA and convert viral RNA into a DNA copy. This viral DNA is then integrated into the host cell's own DNA and from then on is treated by the cell as part of its own (Fig. 4.2).

Retroviruses perform this trick as soon as they get inside a host cell and once integrated they can remain latent for the lifetime of

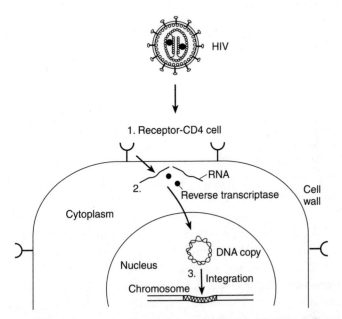

Fig. 4.2 HIV infection of a CD4 T cell. During infection: (1) HIV attaches to CD4 on the cell surface. (2) Virus RNA is released into the cytoplasm where the reverse transcriptase enzyme converts it into DNA. (3) Viral DNA then integrates into the cell's chromosomes and remains there for the lifetime of the cell as a latent infection.

the cell. If the cell divides, the DNA of the virus is automatically reproduced as well. There is no need for the virus to express any of its own proteins, so the cell does not invite immune attack. Although HIV was only discovered in 1983, because it causes AIDS, this retrovirus has been studied so intensively that we know more about it than any other.

Fame and fortune

Like any other field in science, the AIDS field has its fair share of controversy, and the battle that raged over who discovered the AIDS virus exemplifies this. French scientist Francoise Barré-Sinoussi, working in the laboratory of Luc Montagnier at the Pasteur Institute in Paris, was the first to isolate HIV (she called it LAV for 'lymphadenopathy-associated virus') in 1983.[1] Montagnier innocently sent samples of the virus to the well-known American retrovirologist, Robert Gallo at the National Institute of Health (NIH) in Bethesda, Maryland, USA, for further collaborative studies.

In 1984, the Gallo group published a scientific article apparently describing their own isolate of HIV (called HTLV III) which they succeeded in growing in the laboratory in sufficient quantity to develop a blood test for HIV infection and to show that HIV infection causes AIDS.[2] But when both the French and American groups published the genetic sequences of their respective HIVs they turned out to be so similar that it was obvious that they were in fact the same isolate. The alleged implication was that Gallo had in effect 'stolen' the French virus sent to him in 1983, presenting it as a new isolate in his 1984 article. There followed a wrangle, which attracted worldwide attention, over the rights to the diagnostic blood test. So much was at stake, in terms of both finance and prestige, that the dispute went to the very top. It was eventually settled when American President, Ronald Reagan and French Prime Minister, Jacques Chirac agreed to split the credit between the two scientists and the royalties between the two countries.

So that was that until 1989 when there were rumours that a Nobel prize was going to the discoverers of HIV. The argument resurfaced in a special report by the journalist John Crewdson of the *Chicago Tribune* entitled:

'The Great AIDS quest—science under the microscope'.[3]

The article alleged that Gallo had used the French virus to develop his AIDS test, a fact which Gallo was not to concede until 1991. A long enquiry followed which revealed underhand dealing and cover-ups by the American group worthy of the best fictional drama. The congressman in charge of US Government ethics, John Dingell; the director of NIH, Bernadine Healy; and members of the enquiry team itself were all at odds with each other. With leaks, accusations, claims, and counter-claims, the enquiry limped to its final conclusions in 1992.

In a leaked document, Gallo was criticized for 'self-serving behaviour' which reflected a 'disregard for accepted standards of professional and scientific ethics', but he was not found guilty of scientific misconduct. However the document concluded that there had been misconduct in Gallo's group[4], and newspaper headlines like:

'Inquiry concludes data in AIDS article falsified'[5]

pointed the finger at Gallo's assistant of 10 years, Mikulas Popovic, who was accused of lacking respect for truth and accuracy. Some years later these accusations were withdrawn by the Office of Scientific Integrity, so Gallo and Popovic were in some part exonerated. Whatever the truth behind this incident it certainly exemplifies one of the problems facing scientists today. The pressure to get results and to publish them first can only lead to more cases like this in the future.

Relentless spread

When HIV first infects, most people get a mild flu-like illness, or occasionally a more severe disease with fever and swollen glands resembling glandular fever. At this stage, levels of HIV in the blood are very high, the virus spreads throughout the body in CD4 T cells and even lodges in the brain. This infection is brought under control 2–3 weeks later when the immune response is fully operational and the person recovers to remain healthy for an average of 10 years. But the latent virus cannot be eliminated because it has integrated into the DNA of infected cells and is invisibly reproduced each time these cells divide.

But in order to spread between people the virus has to produce new viruses and when this happens the infected CD4 cells make HIV proteins which invite immune attack. At first antibodies and killer T cells control HIV spread in the body. But this virus is adept at evading immune mechanisms, mainly by mutation. Each time the virus reproduces it throws up mutants, some of which will be unrecognized by killer T cells. Then natural selection prevails and since the mutants have a selective advantage over their non-mutated counterparts which are still recognized by T cells, they proliferate wildly. That is until the immune system catches up with them and controls them. But new mutants keep arising and the process is repeated over and over again. Because CD4 T cells that the virus infects and destroys are the very cells which are essential for kick-starting B cells to make antibodies and T cells to kill cells infected with mutant viruses, the body is caught in a downward spiral. Eventually the supply of CD4 cells runs dry and in the resulting immunodeficiency mutant viruses run riot.

During most of the latent period the number of infected CD4 cells in the blood is relatively low. But the blood is not the main home of CD4 cells, it is just the highway they use to get from one place to another. Any snapshot of the infection caught by looking at the blood does not always accurately reflect what is going on elsewhere. The HIV factory is in lymph glands where from an early stage an incredible 100 billion (10^{11}) new viruses are made every day.[6] CD4 T cells normally congregate in lymph glands and interact with other cells there, picking up and passing on messages which are vital for the smooth function of the immune network. Some of these other cells, like macrophages and dendritic cells, also have CD4 molecules on their surface, and can be infected with HIV. When this happens they act as a reservoir of HIV and hand it on to any CD4 cells they come into contact with, so spreading the infection even more widely. One way or another, the virus destroys around 1–2 billion (10^9) CD4 lymphocytes—that is about 30 per cent of the body's total—every day.[6] At first this loss is covered by new cells made in the bone marrow, but eventually demand outstrips supply.

The situation is often compared to a sink with inflowing water from a running tap and outflowing water through a drain. As long as the inflow—that is the new CD4 cells from the bone marrow— equals the outflow of those killed by HIV, normal levels are maintained. But as soon as the outflow exceeds the inflow because of bone marrow exhaustion, the CD4 count falls and immune deficiency supervenes. With this massive daily loss of cells which are essential for immune function, it is only surprising that it takes so long (an average of 10 years) to cause immunodeficiency and AIDS.

The grim reaper

As we have seen, during HIV infection CD4 cells are wiped out in huge numbers, the reinforcements are also infected and destroyed and eventually the loss cannot be replenished. Then the number of CD4 T cells begins to steadily decline. Normally there are around 1200 CD4 T cells per microlitre of blood and this can fall to about 500 without any ill effects. But since these cells play a central role in the body's defences against infection, when their numbers reach the critically low level of 200 things start to go wrong. Opportunistic infections (so called because they take advantage of the situation) occur, and the final stage of infection—AIDS—begins. These opportunistic infections, which are mainly caused by microbes from the patient's own body that would be harmless under normal circumstances, eventually kill the patient.

So it is not always HIV which is ultimately responsible for death from AIDS, but the persistent viruses, bacteria, and fungi which are its bedfellows. And because each person has their own particular combination and levels of microbes in their body, death from HIV takes many different forms. Geographically, the clinical picture of AIDS also varies because the microbes which are common in Africa, for example, are not the same as those common in the West. As the immune system fails, these microbes are no longer subject to the usual strict control and so they cause havoc. Exaggerated, bizarre, and entirely new manifestations of well-known infections occur with each and every system in the body having its own particular problems.

Wasting disease is one of the commonest manifestations of AIDS, particularly in Africa, where it is called 'slim disease'. It has many contributing factors: pain and difficulty in chewing and swallowing food due to ulceration of the mouth and oesophagus is caused by HSV or fungal infections such as thrush; chronic intractable diarrhoea may be due to a galaxy of microbes like giardia, cryptosporidium, entamoeba, shigella, salmonella, CMV, and TB. Any generalized infection like TB may cause severe weight loss and even the long-term antibiotic treatment taken to keep these infections at bay may contribute by disturbing the normal balance of bacteria in the intestine.

Most AIDS patients show signs of brain dysfunction, which range from headache, loss of memory, unsteadiness and tremor, and personality and behavioural changes to acute psychosis, epileptic fits, paralysis, and eventual loss of consciousness. HIV itself infects the brain and may be responsible for some of these symptoms, but other microbes like CMV, herpes simplex virus, toxoplasma, and cryptococcus can all be involved, while Epstein–Barr virus causes tumours in the brain and CMV infection of the retina is a common cause of blindness.

Pneumonia is another frequent terminal illness producing a persistent and distressing cough, shortness of breath, bloody sputum, and chest pain. Here the offending microbes are usually just common bacteria which are normally harmless, but TB and pneumocystis also contribute. Finally, cancers caused by viruses, called 'opportunistic neoplasms', are common in AIDS patients and these are discussed in detail in the next chapter.

This is the awful scenario which most health care workers, scientists, HIV-positive people, and the general public believe is caused by HIV. But a small and vociferous band of scientists, led by the retrovirologist, Dr Peter Duesberg, have always maintained quite categorically that HIV does not cause AIDS. While they agree that both HIV and AIDS exist, they argue that HIV is a harmless virus and the connection between it and the syndrome is fictitious. Duesberg believes in a conspiracy theory regarding the cause of AIDS. Scientists, he says, wish to convict an innocent virus—it is an invention by desperate virologists looking for a reason for their existence. He feels so strongly about this that, in 1996, he wrote a 700-page book, *Inventing the AIDS virus*.

A perilous message

Duesberg claims that, like measles and flu, HIV causes an acute infection, invokes an immune response, and is then eliminated from the body leaving antibodies to protect from further infection. He does concede that on occasions a low level of virus may be left behind, but he argues that this is far too low to cause the disease. He points out that there are many more people who are HIV positive than with AIDS and concludes therefore that the association arises because both HIV and AIDS target the same risk groups, but that HIV infection is much more common than AIDS.

Duesberg also believes that the time between infection with HIV and the onset of AIDS (average 10 years) is too long for the two to be associated. Illness, he says, always strikes within days or weeks of a virus entering the body. He argues that since all viruses reproduce in 8–48 hours, and infected cells produce 100–1000 viruses each time, within 1–2 weeks somewhere around 100 trillion (10^{14}) viruses would be produced. This is enough to infect every cell in the body—so surely it is enough to cause disease. To explain this away, he says, scientists invented the 'latent period' of months or years between infection and disease, and with it the concept of a 'slow virus'. He concludes that 'there are no slow viruses only slow virologists'.[7]

Duesberg has other ideas about the cause of AIDS. 'Sex', he says 'being three billion years old, is not specific to any one group, and is hardly a plausible source for a new disease'. In the West he sees the common health risk as drugs. When AIDS was first reported in the USA in 1981 the 'popper craze' was in full swing in the gay community. 'Poppers' are recreational drugs such as amyl and butyl nitrite taken as inhalant aphrodisiacs. These have toxic as well as the desired effects, and in Duesberg's opinion their prolonged use causes immune deficiency. Heroin, he says, is another AIDS-risk drug, as are antibiotics taken for infections which gay men and intravenous drug users are particularly prone to. Duesberg concludes 'Certainly…drugs consumed at mind-altering concentrations could more easily cause AIDS than a latent, biochemically inactive virus that is present in one out of a thousand T cells'.

For blood transfusion and factor VIII-related AIDS, which most of us believe proves the association with HIV uncontroversially, Duesberg points out that transfusions are not given to healthy people and quotes the statistic that half of all recipients of blood die within a year of the first transfusion. Blood, he declares, is foreign material which overloads an already stressed immune system.

Duesberg even denies that there is an AIDS epidemic in Africa. He calls it the 'myth of an African AIDS epidemic' and maintains that (despite evidence to the contrary) HIV has harmlessly infected Africans for centuries, passing uneventfully from mother to child. This explains why here it is seen equally in men and women, and why so many Africans have antibodies to it. Africans, he says, are dying of the things they have always died of—TB and malnutrition causing slim disease—but now that there is a test for HIV, if it is positive then the death is attributed to AIDS. He concludes, 'whatever does cause early death among third world populations, nothing appears to be new in Africa'.

Duesberg's arguments are as persuasive as they are cynical; his writing as eloquent as it is misleading. Where there is some truth in what he says, his arguments can be answered by accepting that our knowledge is incomplete. HIV is a new virus and AIDS is a new and complex syndrome. There are many facts that we do not yet know or understand about both. But this does not mean that we should abandon the clear evidence we have which unquestionably links the two.

Fortunately, the recent success of highly active antiretroviral therapy (HAART) (see page 195) has at last caused Duesberg to lose his constituency, but in their heyday his maverick views were championed by some in the British media, particularly the former *Sunday Times* editor, Andrew Neil. They not only had dangerous implications for the safe sex campaign, but also led to what the British retrovirologist Robin Weiss terms 'the denial theory'.[8] HIV-positive people and their families, while struggling to accept the awful truth, were only too willing to believe propaganda telling them that the virus was harmless. Those infected often feel well and so may stop taking the only treatment which will benefit them and could even stop taking the precautions necessary to prevent passing the deadly virus on to others.

Proving the obvious

You would have thought that these days Duesberg's case hardly needs answering. But even now government officials in parts of Africa are using his arguments to sidestep their responsibilities. To make the point the counter-arguments are outlined briefly here. Early in the HIV epidemic, as soon as a blood test for HIV was available, studies began in the USA and UK on cohorts of people with a high risk of infection. At this stage the rate of infection was high and infected and uninfected people with the same risk factors could be compared. Studies on gay men showed that only those with HIV infection later developed AIDS. Similar follow-up studies on babies born to HIV-positive mothers found that a quarter acquired the infection and only this quarter eventually got AIDS.

In Uganda there was virtually no HIV before 1980, but by 1990 the country was at the forefront of the AIDS epidemic in Africa. Epidemiological studies there between 1990 and 1996 showed an

8 per cent prevalence of HIV. Now, in the group of those between 13 and 44 years old, those who are HIV positive are 60 times more likely to die over the next two years than their HIV-negative colleagues, and HIV has caused a decline in life expectancy at birth from 59 to 43 years. These results clearly show that in Africa, as in the West, HIV is a lethal infection which causes AIDS.

Early on, an important question to answer was: does AIDS follow accidental HIV infection? Since this involves people outside the typical high-risk categories for infection, the answer to this would prove or disprove the link between HIV and AIDS once and for all. Two studies showed that the answer to the question is yes. Firstly, the unfortunate accidental inoculation of HIV into several thousands of people with haemophilia was the result of using contaminated factor VIII before the danger was realized. In Britain it was relatively easy to track all those given contaminated factor VIII from the comprehensive haemophiliacs register. The infection arrived in 1979 in HIV-contaminated factor VIII imported from the USA and ended in 1986 when the contamination was discovered and the importation ceased. At the time there were 6278 men with haemophilia living in Britain and 1227 of them became infected with HIV. Up until 1984 the death rate among all severe haemophiliacs was steady at around 8 per 1000 per year. Whereas this rate stayed the same in the HIV-negative group of haemophiliacs, it increased tenfold in HIV-positive patients, reaching 81 per 1000 by 1991–2; most of these were AIDS-related deaths.[9]

Secondly, 3000 health care workers worldwide have been accidentally exposed to HIV; 92 of these have become HIV positive and many have already died of AIDS.[10] Malon Johnson is a case in point. He is a pathologist from Vanderbilt University, Tennessee, USA, who specializes in brain diseases and he is HIV positive. In his book, *Working on a miracle*, he describes in graphic detail how he became HIV infected and his subsequent struggle to beat the virus in his body.[11] It makes compulsive reading.

At 8.00 p.m. one evening in September 1992, Johnson was working late at his microscope when the telephone rang. A colleague told him that an AIDS patient had just died and he decided to do the post-mortem examination straight away. As always he took all the right precautions—protective, hooded jumpsuit; face mask and shield; arm protectors and two pairs of latex gloves—but it was not enough. Surrounded by his kit of lethal weapons—saws, knives, scissors, forceps—he set to work to remove the dead man's brain. While peeling away the skin of the scalp before sawing open the skull, his bloody scalpel slipped and cut deep into his thumb. Again he followed all the prescribed guidelines of the time—a thorough wash, encourage bleeding, and disinfectant. The next day he reported to the clinic and signed up for regular blood testing, but that was it: he was HIV infected. From then on his life became a crusade against his own virus. He left no stone unturned, and when the book ends in 1996 Johnson is healthy and taking HAART. He for one must be enraged by Duesberg and his entourage.

The transmissible spongiform encephalopathies

Epidemics of TSEs (BSE, CJD, Kuru) are covered in Chapter 2; here we ask the question, what causes TSEs? The short answer to this question is, we still do not know, but whatever does cause them is microscopic, infectious, and still unclassified. After decades of searching, no virus-like particles have been found to account for TSEs. Despite this there is still an outside chance that an unconventional type of virus will turn out to be the cause. After all, absence of evidence is not necessarily evidence of absence.

Scrapie in sheep is the best-studied TSE, partly because of its economic importance, but also because it infects mice and hamsters which make convenient laboratory models. After injection of extracts from scrapie-infected sheep brain into mice, the mysterious infectious agent can be traced from blood to lymph glands and spleen where it multiplies. Then it makes its way to the brain, probably carried in the blood, where it causes fatal brain degeneration.

Scrapie was found to be transmissible in the 1930s and in 1954 the term 'slow virus' was coined to describe its causative agent. But so far there is no trace of genetic material associated with the disease and these so-called viruses show some very peculiar properties. They are resistant to high temperatures and doses of radiation that would inactivate the genetic material of a regular virus and they do not invoke any kind of response from the body's immune system—no typical inflammation, no antibodies, and no killer T cells.

Until recently, scientists lined up behind one of two opposing views as to the identity of the scrapie agent. There were those, headed mainly by a team of scientists in Scotland, who firmly stuck to biological principles and maintained that there must be a virus or virus-like agent with conventional genetic material involved. And they backed up their theory with experimental evidence. By passing the scrapie agent through generations of mice, they produced over 20 strains of the agent which behaved differently from each other, but each bred true when passed to further mice. To them this meant that each strain inherited its particular set of characteristics from its parent, and to do that, the Scots argued, the agent must contain some sort of inherited genetic material. The other school of thought, emanating from California, and headed by Professor Stanley Prusiner, says that the scrapie agent is nothing but an infectious protein—a prion (a (sort of) acronym for *pro*teinaceous *in*fectious particle).

The revolutionary prion theory, which breaks all the accepted rules of biology and genetics, was proposed by Prusiner in 1982 with no positive evidence to back it up. However it did explain the failure to find any genetic material from a candidate virus in the brains from infected animal and took account of the unusual physical properties of the agent. The proteinaceous material he was referring to was scrapie-associated fibrils found in the brains of diseased animals, which he now called the prion protein (PrP).

To counter the prion claim, the Scottish group invented the hypothetical 'virino'—a self-replicating molecule of genetic material wrapped in a protective coat made of host protein. This contained the required genetic element and was protected from heat, radiation, and immune attack by the normal host protein.

Soon Prusiner located the gene for PrP, which turned out to be the same gene as the Scottish group had previously found to confer susceptibility to scrapie. Then the prion protein—now called PrP(c), for prion protein (cellular)—was found in every cell in all tissues of the body of healthy as well as diseased animals. So PrP is a normal host protein, although its function still remains a mystery. In scrapie brains there is a rogue form of the PrP(c)—called PrP(sc), for prion protein (scrapie). Mice bred in the Prusiner camp, without the PrP(c) gene, do not develop scrapie after infection, whereas mice with the human PrP(c) gene grafted into their DNA do.

The theory is that a change from PrP(c) to PrP(sc) alters the shape of PrP molecules and thus their physicochemical properties (Fig. 4.3). They become less soluble than PrP(c) and cellular

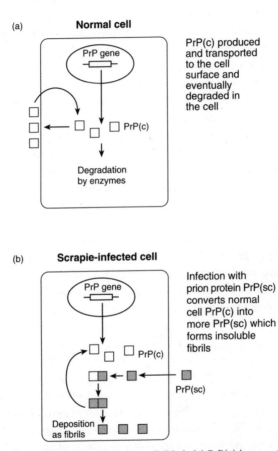

Fig. 4.3 Propagation of scrapie prions—PrP(sc). (a) PrP(c) is a constituent of normal cells which is naturally degraded. (b) PrP(sc) is the mutant infectious form which can convert PrP(c) to more PrP(sc) and is deposited as scrapie-associated fibrils which cause brain degeneration.

enzymes are unable to digest them. So they cannot be degraded and removed, and instead they are deposited in the brain as scrapie-associated fibrils which destroy the tissue. But the most sinister property of PrP(sc) is that it can combine with normal PrP(c) in the brain and convert it into PrP(sc). This forms a chain reaction in which PrP(sc) molecules reproduce themselves. This fact alone, if true, redefines the limits of biology by establishing a reproducing protein as an infectious agent.

Independent scientists in the USA have now confirmed some of these proposed properties of PrP(sc) in the laboratory, and their experiments even suggest that mutations in the PrP gene can reproduce the strain-like differences which Scottish workers had attributed to virino genetics. With all this accumulating evidence in its favour, the prion theory now seems incontrovertible, and in recognition of his ground-breaking work, Prusiner was awarded the Nobel prize in 1997. However, laboratory evidence alone is not proof that PrP(sc) causes the scrapie. Even now there is still an outside chance that some foreign genetic material is lurking inside, and protected by, the rogue protein.

Is it a virus doctor?

Viruses are often used as scapegoats, particularly for diseases where there is no known cause. They have been suggested as the culprit in diseases as diverse as multiple sclerosis, heart disease, and diabetes. In each case the hunt for a specific virus has proved fruitless, but still there are many controversial theories, most of which implicate common viruses in rare diseases. The crucial question is, could a common, well-known virus occasionally change its ways and cause a quite different disease pattern? There are several proven examples of this, three of which are outlined here.

Epstein–Barr virus (EBV) is present in saliva and spreads between people by close contact, like sharing a drinking cup for instance. It usually first infects young children causing no illness and living in the body unnoticed thereafter. This style of infection is almost universal in developing countries, but children in Western societies may miss out on early infection. They then generally get the virus as teenagers or young adults, through kissing. This delayed infection, rather than being silent, often causes glandular fever—aptly named the 'kissing disease'.

Measles virus can also change its spots from its normal pattern to something much more sinister. It usually causes an acute illness with fever and a rash which lasts 1–2 weeks before the virus is eliminated from the body never to infect again. But very rarely, usually when it infects a young child, the virus persists in the brain causing fatal degeneration known as subacute sclerosing panencephalitis, 1–10 years after the initial infection.

Another suggestion is that viruses may do a 'hit and run' job causing damage which only becomes apparent long after they have left the scene of the crime. Autoimmune disease is a good example of this. Common viruses like herpes simplex normally trigger killer T cells directed against viral proteins. But occasionally, quite by chance, the part of the viral protein which these T cells recognize is identical to part of a normal body protein. So T cells generated to fight the virus mistakenly recognize and kill cells expressing this normal protein as well. This is called 'molecular mimicry' and it generates an autoimmune reaction—so called because the immune response (antibodies and/or killer T cells) is misguidedly directed

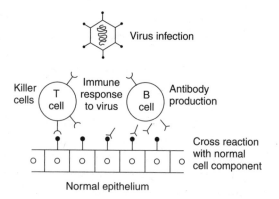

Fig. 4.4 Molecular mimicry. Virus infection induces antiviral T and B cell responses. Killer T cells and antibodies then recognize and target a normal cell component as well as a viral component, causing autoimmune disease.

against self (auto). Here the virus, or indeed any microbe, acts as the trigger to set off the reaction which continues long after it has gone (Fig. 4.4). Herpes stromal keratitis, a disease of the cornea of the eye which can lead to blindness, has been shown in a mouse model to be caused in this way. Because of molecular mimicry between herpes simplex virus and corneal cells, T cells recognizing a particular viral coat protein also target and destroy the cornea.[12]

Multiple sclerosis (MS)

This terrible illness apparently strikes out of the blue, favours young adults (like Jacqueline du Pré, the world-famous British cellist who died of MS in 1987), and can cause devastating and eventual fatal damage to the central nervous system. The disease usually progresses spasmodically with relapses and partial recovery, so that permanent damage to the nervous system accumulates over time. Relapses often seem to follow an acute viral infection, but although virologists have searched high and low, no consistent association with any particular virus can be found. But still many scientists believe that MS must be triggered by a virus.

The features of the disease fit with both the EBV and herpes simplex models outlined above. For one thing, the distribution of MS resembles that of glandular fever. Both diseases target young adults in affluent Western-style societies; both are rare in non-industrialized, tropical countries. However, in addition, MS has an intriguing feature which is very difficult to explain. The incidence of MS varies with latitude, so that it is rare in tropical countries and gets gradually more common towards the North and South Poles; in the UK and the USA it is more common in the North, whereas in Australia it is seen most often in the South. This observation led to the study of those people who have moved from high- to low-incidence areas and vice versa. They found that if young children move, then they adopt the risk of MS found in the new latitude; but when adults move, they retain the risk of their latitude of origin. This suggests that, whatever the predisposing factor for MS is in the high-incidence areas, it is acquired during childhood and then retained for life. Is it a common childhood infectious agent in those areas, perhaps a virus like EBV or measles? This putative virus would infect most people harmlessly as children, but in a few cases, possibly as a result of delayed infection, would cause or predispose to MS in later life.

The brain of MS sufferers contains large numbers of immune cells attacking normal nerve cells—particularly targeting a protein they make called myelin basic protein. This makes some scientists think that MS is an autoimmune disease with molecular mimicry between an unidentified viral protein and myelin basic protein being the cause of the damage. This theory is backed up by experiments using Theiler's virus which infects mice and induces an autoimmune reaction against myelin basic protein, reproducing the symptoms of MS. Over the years many viruses, including measles, EBV, canine distemper virus, and human herpes viruses six and seven, have all been heralded as the cause of MS, but so far none has stood up to rigorous testing.

Chronic fatigue—is it all in the mind?

Chronic fatigue is surprisingly common: 5–10 per cent of visits to family doctors are because of fatigue and when asked specifically, 20–30 per cent of adults admit to always feeling tired. Chronic fatigue syndrome (CFS), defined as unexplained fatigue lasting for six months or more, now affects 250 000 people in the UK, and has recently been accepted as a real medical illness by the Chief Medical Officer. Famous CFS sufferers include Clare Francis (the novelist and yachtswoman) and the daughter of Esther Rantzen (the television presenter).

However, chronic fatigue is not new; it has distinct parallels with a variety of diseases in vogue over the centuries. Up to the end of the eighteenth century these were variously named: bile, melancholia, the vapours (described as 'exhalations developed within the organs…that injure the health'), hypochondria (meaning 'disease located below the sternum'), hysteria (caused by the dry womb of a spinster wandering around the body looking for moisture), and chlorosis (a disease of affluent young women usually cured by marriage and childbirth). In the nineteenth century, CFS-like illnesses included brain fever (the writer Edgar Allan Poe had it) and neurasthenia (exhaustion of the brain and spinal cord as suffered by its discoverer, the American neurologist, George Miller Beard); and the twentieth-century crop encompasses hypoglycaemia (low blood sugar), candida albicans (the fungus which causes thrush), total allergy syndrome, chronic Epstein–Barr virus infection, myalgia encephalomyelitis (ME—inflammation of brain and muscles), and repetitive strain injury. The last was unkindly dubbed 'myalgic compensationitis' when sufferers became eligible for financial compensation.[13] Despite these interesting variations, the best name remains chronic fatigue syndrome because the hallmark is fatigue and no other consistent abnormalities have been found.

CFS re-emerged into the public eye in the 1980s under the name of ME, when a group of scientists claimed to have found a virus which caused unexplained fatigue. The discovery prompted media attention, and the disease was quickly dubbed 'yuppie flu'. Controversy has always surrounded this illness and the media highlighted the opposing views of CFS sufferers on the one hand and doctors on the other. Sufferers mostly feel that they are unfairly labelled as lazy. They believe that CFS has a physical cause, probably a virus infection. In contrast, many doctors feel that the illness has a psychological basis, pointing out the resemblance to depression in its symptoms and in its preponderance of female sufferers.

Despite the media hype, evidence incriminating a virus in CFS is unconvincing. Most patients point to a virus infection at the beginning of their illness, but this coincidence is not surprising since the average person suffers between two and four virus infections every year.

EBV and enteroviruses are the most recent suggested culprits. There is no doubt that glandular fever caused by EBV produces fatigue and in about one in ten people this may persist for more than six months. But in most people with CFS, the onset is not related to EBV infection or to glandular fever. Of course, like 90 per cent of healthy adults, nearly all CFS sufferers are persistently infected with EBV, but the initial infection usually predates their fatigue by many years and latent EBV is controlled normally by their immune systems.

Enteroviruses, particularly Coxsackie virus B4, can infect the heart and skeletal muscles causing myocarditis and myalgia; and in the central nervous system it may cause encephalitis. So clearly this virus is a prime candidate for causing ME. And the studies referred to earlier, carried out in the 1980s, indicated that many CFS patients had evidence of an ongoing Coxsackie infection, with persistent enterovirus in their muscles. However the results of these studies, which were often poorly designed, have not been substantiated by more recent work.

At present we do not know the cause of CFS and given that there are many different causes of tiredness, several may be lumped together under the umbrella of CFS. However, since around half of the sufferers are clinically depressed and their symptoms often improve or cured by antidepressant drugs, in these cases at least there is a psychiatric element perhaps precipitated by stress. Recently a joint working group from the Royal Colleges of Physicians, Psychiatrists, and General Practitioners in the UK reported on CFS[14] and concluded that 'there is no evidence that infections have a primary causal role in the vast majority of cases' and that 'previous personality factors and psychological distress appear to be more important than common viral infection *per se*'.

A rag bag

Several chronic diseases still without an identified cause—including such common afflictions as heart disease, diabetes, and mental illness, as well as the controversial Gulf War syndrome—could all be caused by viruses. But at the moment evidence is fragmentary to say the least. However, since a proven virus cause holds out hope for prevention by vaccination, it is certainly worth searching for viruses which cause these major problems. Cytomegalovirus (CMV) infects over half of the population and sets up a latent life-long infection which, as far as we know, is no problem to healthy people. But now this virus has been implicated in the cause of arterial diseases—atherosclerosis and coronary artery disease—the commonest killers in the Western world. Studies identifying CMV in diseased blood vessels in humans are controversial, but experimental work using a mouse strain of CMV is on firmer ground. Here certainly, the virus grows in cells lining blood vessels causing inflammation and chronic vascular disease. Whether this is also true for humans remains to be seen.

There is no doubt that insulin-dependent diabetes is an autoimmune disease but whether, like herpes simplex-associated corneal keratitis, this is triggered by a virus, is still not proven. Again there is some evidence from mice experimentally infected with Coxsackie virus B4. Here killer T cells destroy the pancreas and wipe out the insulin-producing islet cells embedded in it causing diabetes. But the reason for this mass destruction is not known; possibly molecular mimicry with a viral protein—but possibly not.

Most surprising are recent reports that Borna disease virus may cause psychiatric illnesses such as depression and schizophrenia.

Borna disease virus usually infects horses, although it can also naturally infect a wide variety of animals including sheep, cattle, rabbits, cats, dogs, and ostriches. In horses, the virus first causes abnormal frenzied and aggressive behaviour, and then apathy and withdrawal as the virus persists in the brain. While several studies on Borna disease virus in humans report finding infections there is no agreement as to whether the level of infection in psychiatric patients is higher than that in healthy people. This is partly because infection can be difficult to detect and various different techniques have been used, and also partly because of confusion over the geographical variation in incidence of infection in the general public. Clearly it is important to develop a reliable test for this virus and sort the matter out once and for all.

Finally, the highly controversial 'Gulf War syndrome'. This is not a trivial matter which, as governments clearly hoped, given time would just fade away. During the brief war only 148 of the allied forces died in combat but over 50 000 out of a total of 746 500 British, Canadian, and American troops returned from the Gulf unfit for service. These men and women came down with a bewildering array of symptoms (fatigue, headache, insomnia, tingling of the limbs, breathlessness, and irritability) and the majority of them have remained unable to work.

Rather late in the day the British Government sponsored research into the syndrome, but to sort this one out is a nightmare. In the space of a few weeks troops experienced several major shocks to the system—they were vaccinated against a whole battery of potential biological weapons including anthrax and plague, drilled to respond to poison-gas attacks, inhaled toxic smoke from burning oil wells, witnessed the killing of thousands of Iraqi troops, and withstood SCUD missile attacks. Any one of these high stress-related factors could be the cause of their illness, or might be enough to alter their immunity and allow common viruses to cause unusual symptoms.

There is no clear answer yet but results to date suggest that there is no unique Gulf War syndrome. Most probably, like CFS, it is a multifactorial illness in which stress plays a prime role. Sufferers do not like this conclusion, for everyone it seems prefers a physical cause for their illness. Equally, if true, it does not bode well for preventing such war casualties in the future.

To summarize this chapter, persistent viruses strike up a much more stable relationship with their respective hosts than acute viruses do. They skilfully dodge immune destruction and exploit the host to ensure their own long-term survival. In doing so they generally cause very little harm, but still there can be problems. The most obvious of these is immunosuppression in the host leading to reactivation of the virus, but there are also more subtle sequelae.

A cell which is persistently infected with a virus carries a set of foreign genes which can be passed on to its daughter cells. Although it may not produce overt disease, this life-long intimacy between virus and cell can influence the destiny of both parties. Some persistent virus infections eventually lead to cancer and these viruses are discussed in the next chapter.

Chapter 5

Viruses and cancer

When their first child, Jessica, is born apparently fit and healthy, Dawn and Peter French are delighted. But during the first few weeks at home Jessica is listless; she isn't interested in her feeds and doesn't gain weight. So her parents take her to the family doctor who examines her and thinks she can hear a heart murmur. She refers Jessica to the heart specialist at the local hospital where they find a congenital heart defect. After a battery of tests, the doctor tells Dawn and Peter that it is not possible to correct the defect surgically. However, he thinks that if Jessica leads a sheltered life avoiding exercise and contact with infections, she should be alright for many years.

All goes well for a while but, at a routine check-up at the age of four, the doctor finds that Jessica's heart is failing. He tells Dawn and Peter that there is only one option left—a heart transplant. Jessica is put on the waiting list and by the time a suitable heart is found for her she is perilously close to death.

Jessica comes through the operation successfully, although afterwards the doctor has trouble preventing her immune system rejecting her new heart. Eventually she leaves hospital taking drugs to control the rejection. For a month or two she is well and able to do a lot more now she has her new heart than she could before. Dawn and Peter are optimistic that she has a long and active life ahead of her. Then three months after the transplant her neck glands start to swell up. They are not painful but Dawn takes her back to the hospital for a check-up. The doctor is obviously concerned and a piece of gland is removed under local anaesthetic for examination under the microscope. Cancer is diagnosed and because it has already spread there is very little that the doctors can do. They try chemotherapy but three months later Jessica dies despite the fact that her new heart is still in good working order.

On the face of it this seems like just bad luck—first a congenital heart condition, then an unrelated cancer. But they are not unrelated; they are linked by the transplant or, to be more accurate, by the drugs which Jessica took to prevent the foreign heart from being rejected. These drugs paralyse the immune system and in doing so sometimes allow persistent viruses, including tumour viruses, to get the upper hand. In this chapter we explore the subtle ways in which these viruses promote their own survival—strategies which sometimes inadvertently lead to cancer.

Cancer the crab

One in three people suffer from cancer and, despite all the recent advances in knowledge and treatment, the majority of people who have it die of it. Cancer is the result of a single cell in the body going crazy and growing unchecked until it has produced a whole mass of similar cells—a tumour. This can happen in any organ of the body, in people of any age, and in any country: no one is immune. Paradoxically, although it is common to meet people with cancer, since each person is made up of around 10^{14} cells, at the level of individual cells cancer is extremely rare. Given that one in three people get cancer, the chance of one cell developing into a cancer cell is 1 in 3×10^{14} (that is, 1 in 300 000 000 000 000).

Cancer is not a single disease with a single cause. Rather it is the final result of damage to a cell which may be caused by a multitude of different factors. Most cancers seem to appear for no reason in perfectly healthy people and the cause is usually unknown. So far researchers have found evidence of viruses in 10–20 per cent of cancers worldwide, but other cancers are linked to people's habits, lifestyle, or occupation. Everyone knows that smoking predisposes to lung cancer and that excessive sunlight can cause skin cancer. However, there are many other associations. As exemplified in survivors of the atomic bombing of Nagasaki and Hiroshima in 1945, exposure to radiation increases the chances of almost all forms of cancer. More recently, in the aftermath of the Chernobyl nuclear power plant disaster, a distressingly high number of children developed cancer—particularly of the thyroid gland. Cabinet-makers, who used to inhale the fine dust generated by sawing hard woods, had an increased risk of cancer of the nasal sinuses, and people working with asbestos risked developing a mesothelioma (a cancer of the lining of the lung).

Of mice, chicken, and rabbits

Long before viruses were actually seen under the electron microscope, animal experiments suggested that filterable agents were related to cancer. In 1908, two Danish scientists, Wilhelm Ellermann and Oluf Bang, found that leukaemia in chickens was infectious. By making an extract of leukaemia cells and passing it through filters which trapped even the smallest bacteria, they showed that this filtrate could transmit leukaemia to healthy chickens. But in those days leukaemia was not regarded as a form of cancer and the discovery passed virtually unnoticed. However, in 1911, Peyton Rous at The Rockefeller Institute in New York, USA, also working on chickens, did a similar experiment using an extract of a solid tumour. His tumour came from the right breast of a barred Plymouth Rock hen belonging to a Long Island chicken farmer. The unfortunate farmer took his prized hen to Rous in the hope of a cure, but Rous promptly killed it off and used the tumour for his own research. He injected the extract into healthy chickens and produced the same type of tumour.

For a long time the implications of this work were not accepted by other scientists, particularly those engaged in cancer research. They felt sure that toxic chemicals were the cause of all cancers and simply could not believe that cancer could be caused by a virus because it is clearly not infectious in the same way as diseases like measles and chickenpox. So Rous had a hard struggle for recognition against the closed minds of his contemporaries and eventually turned to experimenting with toxic chemicals himself. But slowly more evidence accumulated and, over 50 years later in 1966, he was awarded a Nobel prize for his work.

In the 1930s, Richard Shope, also working at the Rockefeller Institute in New York, heard from a game-hunter that there were many rabbits in the State of Iowa with large, warty skin tumours. Having obtained some of these rabbits, he managed to transmit the warts to healthy wild rabbits by painting a filtered tumour extract onto their skin, and when he applied this extract to the skin of domestic rabbits, some of the warts grew into spreading skin cancers.

The next breakthrough came in 1936, thanks to John Bittner who was studying breast cancer in mice at the Jackson Laboratory in Maine, USA. He used two strains of mice—one with a high incidence of breast cancer and the other with a low incidence—to carry out a simple but convincing experiment. First, he took newborn mice from the two strains away from their natural mothers and put the offspring of high tumour-incidence mothers to be suckled by low-incidence foster mothers and vice versa. Then he sat back and waited to see if the incidence of cancer in the fostered mice differed from those left with their natural mothers. This showed conclusively that the high tumour incidence was caused by something transmitted in milk from the high-incidence foster mother and not, as was generally believed at the time, by an inherited genetic susceptibility from the biological mother. That 'something' in milk turned out to be a virus. With these discoveries the scientific community finally sat up and took notice—the era of tumour virology had begun.

Slow progress

After the successes in animal tumour virus research, years of disappointment lay ahead for human tumour virologists. Although there are many human cancers, both solid tumours and leukaemias, that resemble cancers caused by viruses in chickens, mice, and rabbits, no human tumour viruses could be found. Many scientists started to doubt whether any human cancers were caused by viruses, while others turned to epidemiologists for help.

Epidemiologists are scientific detectives who study whole populations, looking for clues which link cause and effect (they were the first to demonstrate the link between smoking and lung cancer). If a tumour is caused by a virus then it should be infectious and spread from one person to another by one of the routes which other viruses commonly use (outlined in Chapter 1). So, depending on its method of spread, the tumour might, for instance, show geographical variation, being common in areas where the virus is common and rare where the virus is rare. By the same token, tumours might be more common among family members, close associates, or people with the same occupations or habits, than in the general population. Also people who are highly prone to virus infections of all kinds because their immune system is defective might be more susceptible to tumours caused by viruses. These were the clues that tumour virologists were looking for in the early 1960s when the following event took place which proved a milestone in the history of virology.

Out of Africa

A virologist, Sir Anthony Epstein, was working on Rous' chicken tumour virus at the Middlesex Hospital in London when, in 1961, he attended a lecture entitled:

'The commonest children's cancer in tropical Africa—a hitherto unrecognised syndrome'.

The lecture was given by a British surgeon, Denis Burkitt, working in Uganda. He described tumours of the jaw in African children which grew rapidly, killing in a matter of months. This was a 'new' type of tumour not seen before in Britain, Europe, or the USA, and therefore was of enormous interest in itself. But even more interesting was that Burkitt had noticed a geographical variation in the tumour incidence; in some Ugandan villages there were many children with tumours whilst in others there were none.

Intrigued, Burkitt set off in 1961 on what he called his 'long safari'—a 10 000-mile tour of East Africa—to map the distribution of the tumour (Fig. 5.1). Accompanied by two medical missionaries and travelling in a second-hand station wagon, he visited over 50 large, small, and bush hospitals, and talked to many hundreds of doctors along the way. He always gave an amusing account of this field trip; at one stop, for example, he found a mortuary attendant whose hobby was to make models out of clay. He often used the corpses from the mortuary as models for busts—presumably because they remained still for longer than live subjects. He had made several models of the heads of children with the jaw tumours which Burkitt had come to investigate and he proudly lined them up for inspection on the concrete pathway outside the mortuary. They were so lifelike that while Burkitt was examining them

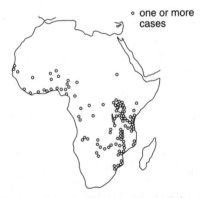

Fig. 5.1 Burkitt's map of the distribution of Burkitt's lymphoma in Africa, plotted after his 'long safari' in 1961. (Reproduced with permission from 'Determining the climatic limitations of children's cancer common in Africa' by D. Burkitt, *BMJ*, **Vol. 2**, pp.1019–23, 1962.)

someone passing by commented on the novel method of treat-ment—burying children up to the neck in concrete.

As a result of his trip Burkitt was able to map out the places in Africa where the tumour occurred. He found that it was restricted to areas where rainfall was above 55 cm per year and the tempera-ture did not fall below 16°C—in other words, areas with a hot and humid tropical climate. This confined the tumour to low-lying areas of central Africa below 5000 feet, and gave it a very similar dis-tribution to holoendemic malaria (non-seasonal malaria which occurs at the same rate all the year round).[1] (Later, the tumour was also found to occur in Papua New Guinea where the same condi-tions prevail.) The reason for the geographical restriction of malaria is the life cycle of the *Anopheles* mosquito which spreads the parasite from one person to another. The larval stage requires water, hence the high rainfall requirement, and adult females do not lay eggs at temperatures below 16°C. In areas of Africa where these requirements are not fulfilled, malaria is seasonal, appearing with the rains and disappearing again in the dry season. Burkitt was familiar with malaria and he put two and two together to suggest that the tumour was caused by an infectious agent which was spread by a mosquito vector. In fact, whereas he was right about the infectious agent—a virus—and about the association with malaria, the virus is not spread by mosquitos.

Listening to Burkitt's presentation, Epstein was fired with enthu-siasm because he realized that, of any tumour he knew, this was the one most likely to be caused by a virus. He immediately arranged with Burkitt for tumour samples to be flown from Africa to his lab-oratory in London. This is not easy because tumour cells are very fragile and start to die as soon as they are removed from the body, so they must be kept alive in a liquid culture medium which supplies all the nutrients they need. They also have to be kept sterile and this was difficult under the primitive conditions in which Burkitt had to work. If the tumour was inadvertently contaminated with bacteria then these would grow in the culture medium and exhaust the nutrients, resulting in the death of the tumour cells.

Despite all the hazards, Epstein got his samples; but disappointment was in store. For two years he and his co-workers tried to isolate a virus by all the techniques known at the time. They searched tumour material under the electron microscope for virus particles, but found none. They put small pieces of tumour into a culture of liquid growth medium closely resembling body fluids to try to persuade the cells to grow, but had no success there either. Most people would have given up long before two years, but Epstein persevered because, he said, 'it just had to be right'.

In 1964, when the breakthrough eventually came it was, as it so often is, quite by chance. One of the valuable tumour samples got held up in transit and finally arrived at the laboratory late on a Friday afternoon. When Epstein peered into the bottle he saw that the medium was cloudy—normally a sure sign of bacterial contamination. It must have been tempting to throw it away there and then, but typically he meticulously took a look at it under a microscope. To his surprise the cloudiness was not caused by bacteria but by lots of tumour cells which had shaken free from the tumour lump during the long journey. The floating cells reminded him of cells from a mouse tumour which he had seen growing in another laboratory, so he put his cells into culture, not as the usual tumour lumps, but as a suspension of single cells. For the first time the cultured cells grew, and once he had enough growing he examined the cells under the electron microscope—at last he found some virus particles.[2] Epstein remembers, 'I simply switched the microscope off, went out and walked round the block two or three times in the snow before I dared come back and look again'.[3]

The reason that Epstein saw the virus in this particular sample and not the many earlier ones was because this was the first one to grow in culture *before* being processed for electron microscopy. Culturing was essential since although all the tumour cells contain the viral DNA, it is in a latent form invisible under the microscope. When in the body, any tumour cell starting to produce virus particles will immediately come under attack from killer T cells, so only when cells are grown outside the body and away from immune defences can whole virus particles be seen.

Epstein went on to show that the virus he had seen was a new type of herpes virus which was consistently found in these African tumours. The tumour, for obvious reasons, became known as Burkitt's lymphoma, and the virus is Epstein–Barr virus (EBV), named after Anthony Epstein and Yvonne Barr (Epstein's research assistant at the time and the one responsible for culturing the tumour cells).

This was the beginning of a life's work for Epstein and his team. They had to find out if the virus actually caused the tumour and then convince others with their experimental results. This was not an easy job; just because a virus is found in a tumour it does not mean that it causes it. It may, for example, have infected the tumour cells after the tumour had grown. It took Epstein many years to persuade some scientists that this virus was the cause of Burkitt's lymphoma, but since those early days EBV has been found in several other types of tumours including nasopharyngeal carcinoma (Fig. 5.2d) and cancer of the lymph glands in people whose immune system is weak (such as Jessica in the story on p. 153). But EBV infection does not always cause tumours. The virus infects most people silently and occasionally causes glandular fever. The discovery of this first human tumour virus renewed the flagging enthusiasm of tumour virologists and opened the way for other important discoveries.

How many more?

Since the discovery of EBV, five more human tumour viruses have been found including viruses which cause common solid tumours such as cancer of the liver and cervix. Like Burkitt's lymphoma, the incidence of all these tumours varies around the world (Fig. 5.2). In total, viruses are involved in one in every ten cancers in men and even more in women; the difference between the sexes being due to cancer of the cervix, one of the commonest cancers in women, which is caused by a papilloma virus.

We met papilloma viruseses which cause harmless warts in Chapter 1, and the devious way they manage to persist in the skin is described in Chapter 4. However, their family contains over 60 different types, and although most are harmless, a few are decidedly unfriendly. In particular, types 16 and 18 can cause anogenital cancers, commonly cancers of the cervix in women and the penis in men. These viruses are spread by sexual contact and the risk of cervical cancer is linked to sexual intercourse from an early age with multiple sexual partners. Conversely, cervical cancer is uncommon in sexually inexperienced women, being particularly rare in nuns. Although seen throughout the world, cervical cancer is most common in the non-industrialized countries of Central and South America, sub-Saharan Africa, and South East Asia.

As we saw in Chapter 1, both hepatitis A and B viruses cause inflammation of the liver. But whereas hepatitis A contaminates food and produces an acute infection usually with rapid recovery, hepatitis B virus is spread by blood contamination and sets up a life-long persistent infection in around 10 per cent of those it infects. This can lead to chronic hepatitis or cirrhosis with eventual liver failure, or, some 20–50 years after the initial infection, to liver cancer.

Hepatitis B virus was discovered by chance in 1964 by Baruch Blumberg working at the National Institute of Health in Bethesda, USA.[4] His work involved a topic quite unrelated to virology—the genetic differences between ethnic and racial groups. He compared blood samples from many different races, and during this he found an unknown protein in the blood of an Australian Aborigine which he called 'Australia antigen'. Several years later he discovered that this antigen was in fact a virus—the cause of 'serum hepatitis'. (At the time this was the name given to hepatitis spread by blood cont-amination or transfusion which had caused major outbreaks of jaundice among troops given contaminated transfusions and vac-cines during the Second World War.)

After Blumberg's discovery a blood test for hepatitis B virus was developed which not only put a stop to the use of contaminated blood but also allowed large-scale population screening. This uncovered an estimated 300 million virus carriers worldwide with particularly high incidences in sub-Saharan Africa and South East Asia. Since liver cancer is among the leading causes of death pre-cisely in these areas (one million die of it every year), the link between virus and cancer was finally made. Thus although hepatitis

Fig. 5.2 Geographical areas of highest incidence of virus-associated tumours: (a) T cell leukaemia linked to HTLV1 infection; (b) liver cancer linked to human hepatitis B (HBV) infection; (c) cervical cancer linked to human papil-loma virus (HPV); (d) Burkitt's lymphoma (BL) and nasopharyngeal carci-noma (NPC) linked to Epstein-Barr virus (EBV).

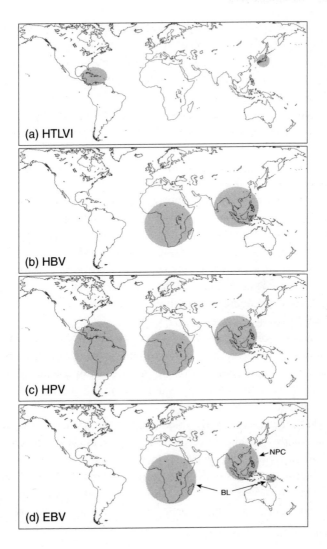

B virus was first discovered in the same year as Epstein discovered EBV, it took a decade to recognize that it was a tumour virus.

After finding that hepatitis B was the main cause of serum hepatitis, it soon became obvious that the latter still occurred despite screening blood for hepatitis B, and therefore there must be another cause, probably an as yet unidentified virus. But the discovery of hepatitis C virus had to await the advent of more sophisticated molecular techniques in the 1980s when its genetic material was identified in the serum of a patient with chronic liver disease.[5]

Hepatitis C virus is mainly spread by contaminated needles and is quite unrelated to hepatitis B virus. However epidemiological surveys show that it too can cause life-long infection and is linked to liver cancer. There is still a great deal to learn about this virus but it seems that infection is very widespread. Around 170 million people worldwide are infected, mostly without knowing it. In the USA alone there are four million carriers and the virus causes 8000–10 000 deaths a year.

Human T cell leukaemia virus-1, known as HTLV1, was the first human retrovirus to be discovered and, although many animal retroviruses cause tumours (those investigated by Ellermann and Bang, Rous, and Bittner were all retroviruses), this is the only human one so far found to do so.

HTLV1, which is spread by blood transfusion, sexual contact, and from mother to child, targets CD4 T cells and establishes a lifelong silent infection. One in 80 HTLV1-infected people eventually develop leukaemia some 10–40 years after first infection. In the Caribbean, where the virus is quite common, it also causes a chronic neurological disease called tropical spastic paraparesis.

Robert Gallo (of HIV fame; see Chapter 3), working at the National Institute of Health in Bethesda, USA, spent many years trying to track down a human leukaemia virus and finally discovered HTLV1 in 1980. The first isolate came from the blood of a young black man who had an aggressive form of T cell leukaemia.[6] Several more isolates followed, all from adults who had similar rapidly fatal leukaemias, often spreading to the skin forming characteristic nodules, and all of whom had links with either the Caribbean or Japan. Gallo's work was published in 1980, but at the same time workers at Kyoto University in Japan found clusters of patients with the same type of T cell leukaemia on the islands of South West Japan. Cells from these leukemias all contained the same type of virus.

A search for healthy HTLV1 carriers turned up a geographical puzzle: 3–10 per cent of healthy people in South West Japan and the Caribbean are positive for the virus, and there are also pockets of infection in sub-Saharan Africa, Eastern Siberia, and Papua New Guinea. Infection in the rest of the world's population is extremely rare. Based on these findings Gallo suggested that the virus originated in Africa, was carried to the Caribbean with the slave trade, and to Japan by sixteenth-century Portuguese traders who travelled via central Africa taking local people and animals with them from there to Japan. We now know that retroviruses carried by African monkeys, particularly chimpanzees and African green monkeys, are 95 per cent identical to HTLV1, and are therefore the likely source of the human infection; this adds credence to Gallo's theory. However, more recent findings of the virus in some American Indians and in the Ainu people of Northern Japan, who are both geographically and racially distinct from the rest of Japan and who had no contact with the Portuguese, suggests a more complicated pattern of spread.

Human herpes virus 8, the latest addition to the list of tumour viruses, was discovered in 1994 when a husband and wife team, epidemiologist, Patrick Moore, and molecular biologist, Yuan Chang from Columbia University, New York, found a completely new virus in Kaposi sarcoma (KS). This is a type of skin cancer which mainly affects blood vessels and causes unsightly purple-coloured patches on the skin and in internal organs. The tumour used to be extremely rare except around the Mediterranean and in East Africa (particularly Uganda), but it has been seen a great deal recently because it is very common in AIDS patients. In the USA and Europe one in every five HIV-positive people develop KS.

Both the geographical restriction and the high incidence in AIDS patients with suppressed immunity suggest that this tumour is caused by a virus. Interestingly though, within the HIV-positive group, KS is 20 times more common in homosexual and bisexual men than in people who acquired the virus through infected blood, heterosexual contact, or transmission from mother to child. This suggests that KS is caused by a virus which spreads by sexual contact and which is common among gay men. Many familiar viruses, including cytomegalovirus, hepatitis B virus, and HIV itself, have been suspected of causing KS in the past, but each one in turn has been eliminated due to lack of incriminatory evidence.

With this information in mind, Chang and Moore decided to look for an entirely new virus in KS tumours. The virus that they eventually found is another herpes virus, quite similar to EBV, but the way it was discovered could not have been more different. They used a highly sophisticated technique called 'representational difference analysis' which simply had not even been thought of when

Epstein and his colleagues were working back in 1964. This technique detects foreign DNA sequences in tissue samples. In this case 'foreign' means viral and the DNA they used was made from a KS tumour. This was compared with normal skin DNA from the same person. Chang and Moore found two DNA molecules which were unique to the tumour material and when they studied the extensive published list of DNA sequences they found just two other DNA sequences which were similar and nothing identical. Both of these sequences came from herpes viruses—herpes virus saimiri (a tumour virus of monkeys) and EBV. Since EBV, as we already know, was the only known human tumour virus in the herpes virus family, Chang and Moore must have been delighted. It meant that they had something similar to two known tumour-causing herpes viruses and therefore there was likely to be another, as yet undiscovered, tumour virus.

Luckily Chang and Moore did not have trouble getting other scientists to take an interest in their discovery and soon their findings were verified. This was such exciting and important work that many scientists around the world were looking at KS material using probes designed from Chang and Moore's unique KS DNA sequence even before the work was published in the American journal *Science* at the end of 1994.[7]

A great deal has happened since then. The virus, now called human herpes virus 8, is consistently found in over 95 per cent of KS tumours of all types (Mediterranean, AIDS-related, and African), although the geographical restriction of the African type has not yet been explained.

The unusual thing about this discovery was that for some time there was no virus. No one had actually seen virus particles, they had only detected the DNA sequences which, because they were not part of the normal human DNA, were assumed to be from a virus. The lack of virus particles in the tumour tissue is familiar from the discovery of EBV and again is due to the latent type of infection that the virus establishes in tumour cells. And again just like EBV, herpes virus particles were seen when KS tumour cells were grown in tissue culture for a while before being looked at under the electron microscope.

Since an absence of virus particles does not exclude a viral cause, the powerful new technique used to find human herpes virus 8 in KS is now being used to hunt for viral sequences in other tumours and leukaemias. This may show that many more cancers are caused by viruses.

The vital link

Discovering a virus in a tumour is just the beginning of the work required to prove that it actually causes the tumour and this may not even be possible because while the direct experiment of simply injecting the virus into the host and waiting to see if a tumour develops can be done in animals, it is clearly not an option in humans. Thus all evidence in humans has to be indirect. The following four questions need to be answered before a virus can finally be said to cause a certain type of tumour.

Is every tumour patient infected?
The first thing to show is that all people with a particular type of tumour are infected with the suspected tumour-causing virus. Testing blood samples for antibodies to the virus as a marker of infection is usually the best way to find out and this means large-scale screening of samples from tumour patients, people thought to

be in high-risk groups, and control groups. Needless to say the answer is often not clear-cut. Rarely, if ever, are all the tumour patients and none of the control group without tumours positive for antibodies, with the high-risk group falling somewhere in between. In the case of human herpes virus 8, early studies do seem to confirm that the virus is more widespread in Africa (but not just Uganda) and in AIDS patients before KS develops.

Antibodies to hepatitis B, which are found in virtually all liver cancer patients, are also detected in 300 million people worldwide of whom only a small proportion—an estimated 1.5 million—will develop the tumour. Similarly, almost all African Burkitt's lymphoma patients are positive for antibodies to EBV—but then so are almost all other African children, along with around 95 per cent of the adult population worldwide. However, with this tumour, levels of antibodies in the blood vary—tumour patients have somewhere around ten times more than people without tumours.

Although these findings show that virtually no uninfected people get the tumours, they do not conclusively prove or disprove the virus/cancer association. Since most of these viruses are extremely widespread, the findings indicate that if viruses do cause tumours then there must be a more complex relationship than the simple equation:

$$\text{virus infection} = \text{cancer}.$$

Is the virus in the tumour cells?

The next thing to do is to demonstrate the genetic material of the virus in all the tumour cells. This is quite easy these days with sensitive molecular probes which can detect the DNA or RNA of the virus under test. However, some scientists suggest that a virus may infect a cell, cause it to become a cancer cell, then leave again, like a thief in the night, leaving no trace. This would be possible if the effect the virus had on the cell was irreversible and inherited by daughter cells when the cell divides. This has never been proven—but then it would be very difficult, if not impossible, to do so.

Does the virus make normal cells grow?

You might expect a virus that can turn a normal cell into a tumour cell to have some effect on these normal cells in the laboratory and this can be tested if the normal cells will grow in culture. With EBV it is easy to demonstrate and the effects are dramatic. The normal counterpart of the cells found in Burkitt's lymphoma are B lymphocytes which usually die after one to two weeks in a culture system. But if they are first infected with EBV and then cultured under identical conditions, they immediately begin to grow and divide, become immortal, and form cell lines which, as far as we know, keep growing forever.

Unfortunately, for other human tumour viruses it is not so easy—either the viruses are difficult to obtain in large amounts or the normal cells required for the experiment do not survive in culture. Both of these problems apply to human herpes virus 8. A recent alternative approach has been to insert individual genes of a virus into cultured cells and look at their effect on cell growth. The E6 and E7 genes of human papilloma virus types 16 and 18, for example, make human epithelial cells grow indefinitely in culture, whereas the same genes from the papilloma viruses which only cause benign warts do not.

Does the virus cause tumours in animals?

If a suspected human tumour virus can be shown to cause the same type of tumour in an animal model, this adds conviction to the suspicion. But demonstrating this can be a problem because many viruses only infect one species—in this case man—and even if they do cross species barriers they usually only infect closely related animals. This may mean working with monkeys which are expensive, difficult to house and handle, and are often protected species. EBV, for example, causes tumours in cotton-topped tamarins, a small South American primate which is an endangered species and can therefore only be used for experimentation if bred in captivity. Similarly, hepatitis B virus will only infect species closely related to man, such as chimps. These days, for a variety of reasons, this type of experiment is actively discouraged.

The next best thing is to see whether related viruses cause similar tumours in the animals which they naturally infect. Ducks and woodchucks, for example, have their own hepatitis viruses which cause liver cancers. In general however, most animals, like most humans, are resistant to tumours caused by viruses they carry naturally.

It may be possible with some or all of the above evidence to convince people that a virus causes a particular tumour, but that does not say anything about *how* it does it. The important question to answer here is: what does a virus do to a normal cell to turn it into a tumour cell? We do not know the whole answer to this yet, but using new molecular techniques we are certainly getting closer to understanding the process.

Loss of control

Cancer cells are totally out of control. They abide by none of the rules which regulate normal cells, but just divide endlessly, regardless of anything else, squashing, invading, and destroying normal tissue as they go. So before we can understand how cancer occurs, we have to resolve the question 'how are normal cells controlled?'. What, for example, makes a cell decide to divide and what tells it to stop again? The skin is the best example to illustrate this point because it fits our bodies perfectly—we are born with a compact outer covering and as we grow, it grows with us. It does not get so stretched that we look like an over-inflated balloon or split when we try to move; neither is it so loose that it hangs in folds, resembling a bloodhound. The number of skin cells increases to cover our growing body and if we make a hole in the covering, cells grow in from the sides until the hole is mended and then they stop growing again. How? What controls them?

Cells respond to an array of chemical messengers or signals from their surroundings which tell them when to divide and when to rest. These signals are released by one cell and received by a series of receptors on the surface of another. After a skin injury, damaged skin cells release chemicals which lock onto the receptors of surrounding cells. This docking mechanism transmits a signal through the cell cytoplasm to the nucleus. The nucleus is the control centre of the cell and it issues instructions for the cell to divide until the breech is mended. Then it tells the cell to stop dividing. The set of genes switched on by signals telling the cell to divide are called cell cycle progression genes or oncogenes. Those which respond to inhibitory signals and stop the cell dividing are called tumour suppressor genes (Fig. 5.3).

Fig. 5.3 Control of cell division. Whether a cell divides or remains in a resting state is determined by the balance of signals it receives acting on its oncogenes and tumour suppressor genes. When oncogenes are switched on, the cell will divide. When tumour suppressor genes are switched on the cell will stop dividing.

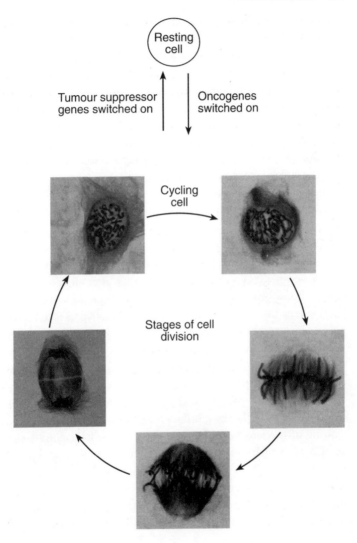

In normal cells the actions of these two sets of genes counterbalance each other, yin–yang fashion, resulting in finely-tuned control. However, if an oncogene is permanently switched on and unresponsive to normal controls, then the cell will divide continuously. Similarly, if a tumour suppressor gene is inactivated and permanently switched off, the effect will be the same—the cell will go on and on dividing. A malfunction in any of the chemical pathways leading from the cell surface receptors to genes in the nucleus may have the same effect. So, there are many ways to turn a normal cell into one which is continuously dividing and unable to stop. This is the beginning of the making of a cancer cell, and these are the complex control mechanisms which tumour viruses usurp for their own purposes.

All tumour viruses interfere with the balance of signals controlling cell division, driving the cell to faster growth. But they do not do this because they want to cause tumours; after all, if a tumour kills the host, the virus becomes a homeless refugee until it can find another host to live in. But since a virus's ultimate goal is to reproduce its own genetic material and for this it needs access to enzymes which the cell uses when copying its own DNA during cell division, many viruses have evolved a variety of clever tricks to make cells divide. They can influence almost any step in the chemical pathways from outside the cell to the genes in the nucleus. And the bigger the virus is, the more influential it can be.

Some of the small retroviruses, like chicken leukaemia virus, actually carry their own viral oncogene. Once their genetic material is converted into DNA and becomes part of the cell DNA, this oncogene behaves just like a cellular gene, forcing the cell to divide. But this direct approach is rare and the human retrovirus HTLV1 uses other strategies. It makes infected cells display receptors on their surface which bind a growth factor responsible for stimulating the infected T cells to divide. EBV carries genes which kick the cell's own oncogenes into action. In the same way, the large pox viruses make their own growth factor receptors which are displayed on the cell surface, acting as locks for growth factor keys just as if they were cellular proteins. But pox viruses do not cause cancer and, fortunately, not all viruses which make cells divide are tumour viruses. If, like pox viruses, induction of cell division allows the virus to complete its life cycle and produce new viruses, then the cell usually dies and a dead cell cannot make a tumour.

It is those viruses which establish a silent or latent infection and remain undetected inside cells for long periods—often years—which are most dangerous. Human papilloma virus, for example, sits in cells in the bottom layer of skin and when the cells divide, the viral DNA divides and sends one copy into each daughter cell. Eventually, the viral DNA may get caught up by one of the cell's chromosomes by mistake and then the virus becomes integrated into the human genome. For papilloma virus this is an essential step on the way to tumour formation.

The kiss of death

Like secret agents who carry poison capsules to be used in desperate situations, cells have a fail-safe, self-destruct mechanism called programmed cell death or apoptosis. The word 'apoptosis', meaning 'the falling of leaves', was coined by a group of pathologists working at the University of Aberdeen and headed by Sir Alastair Currie. They were the first to recognize this distinctive form of cell death.[8]

Apoptosis probably originally evolved to rid the body of harmful viruses and certainly death of a virus-infected cell before it has time to produce thousands of new viruses can prevent infection from spreading throughout the body. Apoptosis may also be triggered if cell division gets out of control. For instance, if a cell receives signals telling it to divide when it is unable to do so because the amino acid building blocks required to make new proteins are not available, then it will die by apoptosis. Alternatively, some tumour suppressor genes activate apoptosis when the DNA of a cell is so badly damaged that it would be dangerous to the animal as a whole to let it divide. The suicide programme activates cellular enzymes which chop up DNA and causing instant cell death. In just 30 minutes the cell is irrevocably damaged and rapidly gobbled up by a passing macrophage.

Over the millions of years that they have been in contact with their respective hosts, both tumour and non-tumour-causing viruses have evolved ways of persuading cells to resist apoptosis. Like cell division, apoptosis has many positive (death or suicide) and negative (survival) control genes whose actions are delicately balanced. And once again, viruses have sophisticated ways of influencing them. Some, like EBV, undermine the cell death programme by carrying their own cell survival gene or by turning on those in the cell. Others, like adenovirus, inhibit the cell's suicide genes. Whichever way they approach it, the outcome is the same—infected cells stay alive long enough to support a productive or latent infection. This can be a crucial event in the evolution of a tumour.

The great escape

Despite their name, most tumour viruses establish lifelong infections in normal people without causing any harm. This is partly because they are so efficiently controlled by the host's immune system that, as we saw in Chapter 4, persistent viruses have to adopt a really low profile in order to remain on board. So it is not surprising that tumours caused by viruses are more common in people whose immune system is not working properly than in the general population. There has always been a small group of people whose immune system is defective because of an inherited genetic disorder, but recently, numbers of immunocompromised people have increased enormously because of the success of organ transplantation, cancer treatment, and also the AIDS epidemic.

Killer T cells circulate around the highways and byways of the body looking for trouble. They recognize and accept anything which belongs and eliminate anything, such as a virus-infected cell, which is foreign. To a killer T cell a transplanted organ, however well-matched to its new host (unless it comes from an identical twin), is foreign and therefore flagged for destruction. To counteract this, after a transplant people take drugs to suppress their killer T cells. The drugs, however, cannot differentiate between T cells which kill the transplanted organ and which therefore need to be inactivated and those which fight infections which ought to be functioning normally. So, people with transplants are more prone to infections since viruses can escape from immune control. Their body is caught in a balancing act between infection and destruction of their new organ.

Transplant recipients have particular problems with persistent viruses which are either already lodging in their body before the transplant or are actually delivered from the donor in the transplanted organ. Some of these—like EBV, human papilloma virus and human herpes virus 8—are potential tumour viruses, so when the dose of immunosuppressive drugs is increased to protect the transplant, virus immunity is reduced and the risk of tumour development rises. This is exactly what happened in the imaginary story of Jessica French at the beginning of this chapter, after she had her heart transplant. In her case the virus implicated was EBV and the tumour was a lymphoma.

Step by step

It is no wonder that the early cancer researchers did not believe that viruses cause cancer because, on the face of it, virus-associated tumours do not resemble classical infectious diseases. They were right to say that cancer does not generally occur in epidemics and is not obviously infectious. Also, most tumour viruses are widespread in the population and reside quite harmlessly in their hosts for life. Clearly there are many more infected people than there are those with the associated cancer, because common viruses are associated with rare cancers. To explain this, the theory of multistep progression from a normal cell to a cancer cell came into being. This states that not just one, but many abnormalities are needed before a normal cell becomes a cancer cell. Most of these defects are genetic accidents affecting the cell's DNA and having the effect of releasing the cell, little by little, from normal control mechanisms until finally, when all the changes are added together, the result is a completely anarchic cancer cell.

In over half of all cancers, regardless of the type of cell involved, destruction of a tumour suppressor gene is one of these genetic accidents and in virus-associated tumours, infection with a tumour virus is another. This allows foreign genes into a cell and destabilizes its control mechanisms. Although not usually sufficient on its own to drive a cell into uncontrolled growth, virus infection is, nonetheless, an essential link in the chain of events which leads to cancer.

So, the reason why tumour viruses are much more common than their associated cancers is that the chance of all the required abnormalities occurring in one cell is very, very small indeed. And the reason for the geographical restriction of some tumours is often that two or more of the factors required for tumour development coincide in those areas.

EBV, for example, is very common, infecting around 95 per cent of the population throughout the world, but a co-factor for Burkitt's lymphoma development, holoendemic malaria, only occurs in equatorial Africa and Papua New Guinea. So those are the areas where Burkitt's lymphoma occurs. For this particular tumour, a genetic accident involving a chromosome switch is also essential. A piece of chromosome 8, which contains the cell oncogene *c-myc*, accidentally gets reattached (translocated) to chromosome 14, as a result of which *c-myc* is permanently switched on, tipping the balance in favour of cell division (Fig. 5.4). As we discussed earlier, the equation 'tumour virus=cancer' is too simple. Now, for Burkitt's lymphoma, three essential events have been identified and so we can write:

EBV+malaria+*c-myc* deregulation=Burkitt's lymphoma.

Several more abnormalities probably contribute, but these have not been identified. Also, it is not yet known precisely what malaria is doing in this scenario or how the *c-myc* chromosome translocation occurs. But at least a picture is starting to emerge which highlights the complicated processes involved, and this should help in planning prevention and treatment strategies in the future.

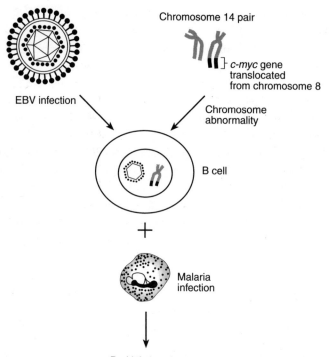

Fig. 5.4 Stepwise evolution of Burkitt's lymphoma. The changes which turn a normal B cell into a Burkitt's lymphoma cell include Epstein–Barr virus infection and a translocation involving the *c-myc* oncogene in the B cell, and chronic malaria in the host. The condition is rapidly fatal if untreated but timely intervention with chemotherapy cures many cases.

This multistep progression towards cancer, with the slow accumulation of abnormalities, each of which is itself rare, not only explains why just a minority of people with a particular virus infection get the associated cancer, but also why there is often such a long interval between initial infection and cancer development. With liver cancer, for example, up to 50 years may elapse between hepatitis B infection and the appearance of cancer and at some time during this period viral DNA becomes integrated into cell DNA. So, among hepatitis B carriers, liver cancer is more common in people who were infected early in life, particularly those infected before, during, or shortly after birth. The high number of hepatitis B carriers in South East Asia means that transmission of virus from mother to child around the time of birth is common and therefore the incidence of liver cancer in this area is very high. Another factor thought to be involved in the development of liver cancer is the ingestion of aflatoxin B1. This toxin is produced by a mould called *Aspergillus flavis* which grows in grain stored in damp conditions. In animals it is one of the most powerful cancer-causing chemicals known.

Hope for a cure?

Finding that a virus is involved in the growth of a tumour opens up several potential new treatment possibilities, most of which involve manipulation of the immune system. The most obvious and straightforward thing to do is to prevent virus infection in the first place. Then, if virus infection is an essential step in development of the tumour, the tumour will not develop. This strategy has been extremely successful in preventing Marek's disease, a tumour of chickens caused by a herpes virus, which used to devastate domestic flocks.

Jozsef Marek, a Hungarian pathologist, first described this disease in 1907 and although he suspected that it was caused by an infectious agent he was never able to prove it. Later, many people tried to transmit the disease from one chicken to another, but the results were never convincing because although their test birds came down with tumours, so did some of their supposedly uninfected control birds.

It was not until 1960, when Peter Biggs started working on the disease at Houghton Poultry Station in Huntingdonshire, that carefully controlled experiments with groups of isolated birds proved that Marek's could be passed from one bird to another. The unusual thing was that the disease could only be transmitted with whole tumour cells—not with a tumour extract. However hard Biggs tried he could not make a cell-free tumour extract which was infectious. So critics said that he was just transplanting tumour cells from one bird to another and there was no need to invoke a virus. In answer to this Biggs devised an ingenious set of experiments in which he transferred tumour cells of male birds to female recipients and vice versa. Then he studied the chromosomes of the resulting tumours to see if they were XX (female) or XY (male). In almost all cases the new tumours were the same sex as the recipient and different from that of the injected tumour cells. This silenced his critics by proving that a transmissible agent released by the tumour cells to infect the recipient bird was the vital ingredient. It also convinced him that a virus was involved, even though he could not separate it from the cells it infected. In 1967, workers in Biggs' laboratory finally succeeded in infecting chicken kidney cells in tissue culture and these produced sufficient virus for study under the electron microscope and to identify a herpes virus.[9]

We now know that in a natural infection Marek's disease virus is produced in large quantities by cells lining feather follicles. The virus is highly infectious, spreading from one bird to another in infected bedding and by inhalation. So the virus sweeps rapidly through flocks and it was this that led to the infected control birds in Marek's early transmission experiments.

Following the discovery of this new herpes virus, Biggs had the task of proving that his virus actually caused Marek's disease and was not just a passenger virus in the tumour cells. At the same time, work began to develop a vaccine and here the story took an unusual turn. A researcher in Biggs' laboratory who was involved in making the vaccine, left, taking some virus with him. He promptly set up a company and manufactured a vaccine of his own. By 1971 his vaccine was on the market and was spectacularly successful in preventing Marek's disease.[10] This rather ungentlemanly act, which shocked the scientific community of the time, made his fortune but, despite holding the patent on the vaccine, Houghton Poultry Station, where the virus was isolated, did not benefit financially.

At first, the prospect of preventing hepatitis B infection by means of a vaccine looked bleak. Although a safe and effective vaccine had been available since 1982, at $100 per course it was so expensive that it could only be used in the West. It was beyond the means of governments and international organizations like the WHO to provide it to countries in sub-Saharan Africa and the Far East, where it was most needed. So an international task force of influential scientists set about pressurizing vaccine manufacturers to reduce the price, convincing governments of the seriousness of the problem and persuading international organizations to provide funds.[11] Thanks to their efforts hepatitis B vaccination programmes are in place the world over and the WHO aims to completely eliminate the virus within the foreseeable future.

A vaccine against papilloma viruses 16 and 18 is being developed and within the next decade it is quite possible that children will be immunized to prevent genital cancer occurring maybe 30–50 years later. This would have a major impact on health care since cancer of the cervix is one of the commonest fatal cancers in women. However, for the present we are faced with people who already have cancer and require urgent treatment. Would vaccination be any use at this stage?

Patients with tumours caused by viruses such as hepatitis B and papilloma viruses have a normally functioning immune system, but still the viruses manage to avoid recognition by killer T cells, probably because the tumour cells are just not displaying enough viral peptides. In this case it may be possible to boost the patient's own immunity enough to tip the balance and persuade killer T cells to destroy the tumour. This is the rationale behind ongoing trials where patients with cancer of the cervix are immunized with viruses which have been genetically engineered not to grow in the body but to produce high levels of viral proteins known to be present only at low levels on tumour cells. The hope is that this will boost the T cells enough to recognize and kill tumour cells.[12]

In other cases, as we have already discussed, virus-associated tumours occur in people with weakened immunity and here tumour cells may display viral proteins which in a normal person would act as the trigger for killer T cells. So all that has to be done is to restore the failing immune system and this is sometimes possible in people with organ transplants by reducing the dose of immunosuppressive drugs and allowing T cells to act normally again. However, doctors often find themselves in a catch 22 situation—give too much immunosuppression and allow the tumour to grow, or give too little so that the tumour may shrivel and die, but the vital transplanted heart, lung, or liver is lost. The only organ which can be lost without fatal consequences is a kidney and here the patient and doctor may decide that it is worth stopping the drugs completely and sacrificing the transplanted kidney to get rid of the tumour.

These days, clones of killer T cells which only recognize and kill EBV-infected cells can be grown in the laboratory. A group of doctors at St Jude's Hospital in Memphis, Tennessee, pioneered the use of these to treat bone marrow transplant patients with tumours caused by EBV,[13] but at the moment this process is too time-consuming and expensive to use for routine treatment. However, killer T cells can be kept in the freezer and so the time may come when 'spare part' surgery is commonplace and each patient will have their T cells, that recognize all the troublesome viruses, stored in the freezer before the transplant operation takes place to be used to counteract infection at a later date.

Since the discovery of the first human tumour virus (EBV) in 1964, five more viruses have been linked to human cancer and there are probably several others out there waiting to be discovered. Although in most cases virus infection is only one of the several 'hits' required for a normal cell to turn into a cancer cell, it can act as a target for tumour therapy. Many new treatments are being tried out and these are bringing hope to a field of medicine in which great strides in basic science are just on the verge of being translated into help for the patient. The next chapter looks at the other methods of attack we have at our disposal to use against viruses.

Chapter 6

Searching for a cure

Nancy Harrison is a GP in a busy university student health clinic. Every morning on her way into her consulting room she passes rows of students waiting to see her with a variety of complaints ranging from coughs and colds to exam stress and depression. This Monday morning is no exception. The first two she sees are remarkably similar.

John and Peter are first-year biologists and know each other well. They both have sore throats and are feeling tired and off colour. They are feverish with enlarged tender neck glands. Dr Harrison treats them both in the same way—the way she has learnt from experience will minimize their visits to the doctor. She takes their temperatures and examines their throats then tells them that they have an infection. At their age, it is most likely to be glandular fever, so she takes a blood sample to test for it. The results will be ready in 24 hours so she asks them to come back in two days' time. In the meantime she gives them both a prescription for penicillin, tells them to take aspirin gargles for the sore throat and to rest until they feel better.

All very routine, but the test results are quite surprising. Although Peter and John's symptoms are so similar, they are actually caused by quite different microbes. Peter tested positive for glandular fever but John did not. John appears in the surgery on Wednesday already feeling quite a bit better; the penicillin is clearly working. Peter does not turn up, so Dr Harrison phones him. His girlfriend Karen answers and says he is too ill to go to the surgery.

Dr Harrison goes to see Peter on Wednesday afternoon and finds him in bed. He still has a fever and sore throat but mostly he feels so completely exhausted that even sitting up is too much for him. He is relying on Karen to bring him drinks and he cannot manage to eat anything. Dr Harrison is not optimistic of a quick recovery and tells Peter that there is no effective treatment for his infection; it will just have to run its natural course—usually 3–4 weeks. Even after that time, people often experience a period of fatigue which makes getting back to normal a slow process. She gives him a painkiller for his throat, tells him to keep drinking, and promises to be in touch in a day or two. When she gets back to the surgery Dr Harrison sends a note to Peter's Director of Studies telling him that Peter has glandular fever.

By the end of the week John has completely recovered and is back at his classes. Peter, on the other hand, is still in bed and feeling exhausted. But the painkillers are working and so he is managing to eat a bit. It takes another week before he is able to get up and dress, but still he is worn out by the effort. The next week he tries to attend classes, but after an hour or two he is unable to concentrate and he begins to worry about his work.

Two weeks later the end-of-term exams are looming and most students are working in the evenings, but it is all Peter can do to get through the day. When he gets home in the evening he just falls into bed. He goes to see his Director of Studies who is reassuring. Glandular fever is common among the students and is well-known for causing prolonged fatigue; he must just do his best in the exams and his illness will be taken into account when the papers are marked.

So four weeks after the initial visit to Dr Harrison, when John has already forgotten about his infection and is fully prepared for the exams, Peter is still struggling with fatigue in the evenings and a lack of concentration during the day. Their contrasting experiences have come about because, whereas John had a bacterial throat infection (probably a streptococcus), Peter was infected with a virus (Epstein–Barr virus). Antibiotics like penicillin kill off bacteria but there is no such effective treatment for the majority of virus infections.

Is cleanliness next to godliness?

The Scottish bacteriologist, Alexander Fleming (1888–1955), was interested in the body's resistance to bacterial infection. He worked on *Staphylococcus*, the bacterium which causes boils and abscesses, which he grew on agar—a jelly-like nutrient set in shallow round dishes. And since he was notoriously untidy, his laboratory at St Mary's Hospital in London was always full of these culture plates which he left hanging around for several weeks. In August 1928 he returned from holiday to be greeted by a pile of neglected staphylo-coccal plates. Before throwing them away he examined them all carefully, and selected one that interested him. It was contaminated with mould—probably not unusual given the conditions he worked under—but what caught his eye was an area of agar surrounding the mould where the staphylococci colonies had disappeared. Interested, Fleming grew the mould in a culture broth and later identified it as a new fungus—*Penicillium notatum*. Much more importantly, he found that the fungus produced a substance which killed bacteria while leaving human cells alive. Fleming realized that this might be used to treat bacterial infections and he tried to purify the active ingredient from the mould. But, with no success and little encouragement from his superiors, he soon dropped the project.

It was some 10 years later that Howard Florey and his colleagues, working in Oxford, finally purified penicillin, and in 1941 a local Oxford policeman received the first dose. He was suffering from staphylococcal septicaemia—a blood infection which at that time was invariably fatal. He improved with the treatment, but unfortunately stocks ran out before he was cured and he subsequently died. By 1944 mass production of penicillin was under way just in time to save the lives of many Second World War victims with infected wounds.

Fleming's unique combination of untidiness and inquisitiveness which led to the discovery of penicillin, heralded the antibiotic era. For the first time, previously fatal infectious diseases could be cured and the effect on death rates was dramatic. The next 20 years saw an explosion in antibiotic production; there are now hundreds of natural and synthetic compounds active against virtually all types of bacteria. In theory, bacteria should not stand a chance against this array of chemical weapons, but in reality they are just as prolific as ever. This is because bacteria multiply so fast (they can double their numbers every 20 minutes) that they frequently throw up mutants. If resistant mutants appear during antibiotic treatment, they have a selective advantage and soon take over.

With the widespread use and abuse of antibiotics over the last 50 years, we have an increasing problem of bacterial resistance and important pathogens like TB, salmonella, and staphylococcus now show multidrug resistance. Alarm bells are ringing and some are predicting the end of the antibiotic era.

The antiviral era

Antibiotic means 'destructive of life' and penicillin, along with its later synthetic derivatives kills bacteria by inhibiting cell wall synthesis. Since the structure of a bacterial cell wall is unique, antibiotics are selective for bacteria, leaving other cells relatively unharmed.

Viruses are not cells. When outside a cell they are particles and when they infect a cell they in essence become part of it. So antibiotics have as little effect on viruses as they do on cells.

Because viral growth is so intimately involved with cell growth, until recently many scientists thought that anything which disrupts one would necessarily disrupt the other and so selective killing of viruses was an impossible goal. However, over the last 20 years this pessimism has been proved wrong; several safe antiviral drugs are now on the market and we are poised at a point equivalent to the dawn of the antibiotic era 50 years ago.

Antivirals generally act by targeting proteins which are unique to the virus and essential for its growth. At first these drugs were discovered by chance, often during the search for new cancer treatments, but now that at least some of the molecular pathways exploited by viruses have been unravelled, rational drug design can target individual viral proteins.

Acyclovir was the first antiviral available. It is active against most herpes viruses and is particularly useful for oral and genital herpes (caused by herpes simplex virus) and shingles (caused by varicella zoster virus). This drug is so safe that people at particular risk can take it over long periods to prevent unpleasant recurrences of these herpes infections. But long-term treatment can produce mutant herpes viruses which are resistant to acyclovir in a similar way as to bacteria are resistant to antibiotics. Fortunately, so far the mutants have been weaker than their non-mutated counterpart and have not spread in the general population. At present only about 1 in 10 people with recurrent genital herpes in the USA take acyclovir, but by using mathematical models to predict the spread of drug-resistant mutants, scientists calculate that even if half of those affected took the drug still five new infections would be prevented for each case of drug resistance.[1]

Drugs are also available for flu virus infection, but despite being safe they are not often used. This is probably because it is difficult to know when and how to use them. They are about as good at preventing flu as the vaccine is, so during an epidemic they could be given to the elderly and chronically sick. Alternatively, they could be taken as soon as symptoms begin, when they would reduce the length of illness by about one day. This would have enormous financial benefit to employers by reducing absence from work, but it is not particularly attractive to doctors who continue to vaccinate their high-risk patients.

Who is in the driving seat?

At the 1993 International AIDS Conference, the outlook for a cure for HIV infection was bleak and the mood was gloomy, but by 1996 the mood had completely changed. By then optimism was so high that some reporters predicted the imminent end of the AIDS epidemic. The buzz-word was 'triple therapy', and for the first time since 1983, HIV infection seemed to be treatable and the death rate from AIDS in the West was falling. This remarkable revolution came about when AIDS doctors, taking a leaf out of the cancer doctors' book, began treating HIV-positive people with combinations of two or three antivirals at the same time, rather than with a single drug.

HIV has two enzymes essential for its life cycle—protease which ensures efficient infection and growth and reverse transcriptase (RT) which turns its RNA into DNA so that it can integrate into host cell chromosomes. Both these enzymes are now targets for antivirals which knock them out so that the virus can no longer function.

The first drugs to come on line were directed against RT and although they had a beneficial effect when first taken HIV quickly mutated to a drug-resistant form making further treatment useless. Now there are many drugs active against HIV. The key is to use a drug combination which efficiently inhibits viral growth early on in the disease. As long as the drugs work, the virus cannot grow and cannot produce mutants which evade immunity. So the infection remains stable. Therapeutic trials show that a combination of three drugs, all taken at once and including at least one protease inhibitor and one RT inhibitor, gives the best insurance against disease progression. Such drug combinations are commonly known as HAART (Highly Active Anti-Retroviral Therapy).

Monitoring viral load, that is the amount of HIV in the blood, tells doctors how effectively the drugs have the infection under control and levels of CD4 cells reflect how the immune system is coping. If viral load begins to rise and CD4 counts fall then it is time to change to another drug combination. The euphoria of 1996 came about when doctors reported many patients on combination therapy with undetectable viral loads. So the possibility of complete eradication of HIV from the body and a cure for AIDS was mooted. In retrospect this was decidedly premature—it will not be that easy to rid the body of a virus which integrates into host cell chromosomes. There are many places in the body, particularly the lymph glands and the brain, where HIV can hide and even in blood, new, more sensitive assays can detect remaining virus. However, it is certainly true that progress of HIV infection can be halted by HAART and there are now more long-term survivors with HIV infection than ever before.

Still, a significant number of HIV-positive people do progress to AIDS; so many in fact that some doctors think that we may only be buying time with HAART. Failure is sometimes because of viral escape mutants; here people simply run out of drug options. But in other cases it is non-compliance rather than mutation which is the problem. These people are just not taking the drugs, either because of nasty side-effects or because they have so many tablets to take every day that it becomes impossible to keep up with every regimen. The simplest drug regime for HIV involves taking 10 tablets a day and full triple therapy may be up to 30. In addition to this, many people have to take drugs to prevent opportunistic infections, so up to 40 tablets a day, probably all taken at different times and frequencies, is not unusual.

Unpleasant side-effects of HAART include nausea, vomiting, and diarrhoea, but more sinister are the effects long-term use of protease inhibitors can have on the body. Patients are reporting changes in their body shape due to a redistribution of body fat which leaves the extremities (face, limbs, and buttocks) wasted, but with central obesity. This typically produces abdomen and breast enlargement and a 'buffalo hump', that is, a pad of fat at the back of the neck which starts as a nuisance when collars get tight but can end up as disfiguring. Of more concern though is the high blood cholesterol and triglycerides—the fats which predispose to high blood pressure, heart disease, and diabetes.

Clearly HAART is not ideal and more antivirals are urgently needed to add to our arsenal against this most persistent attacker. But despite these concerns HAART is a tremendous advance and is certainly the best treatment for HIV infection available at the moment. It is expensive—around £6000–£7000 a year for each patient—but AIDS doctors argue that this is cost effective because it keeps people well. So, fewer hospital beds are required for AIDS patients and less expensive drug therapy is needed for other opportunistic infections. For example, treatment for one case of CMV retinitis costs around £20 000 a year and the number of new cases at a large STD clinic in London fell from 40 to 4 per year after the introduction of HAART.[2]

But this kind of logic only applies in the West. Their high cost prohibits the widespread use of antivirals in developing countries and this has led to a heated ethical debate. In 1994, a trial in the USA and France showed that intensive treatment with the antiviral drug zidovudine (AZT), given to HIV-infected pregnant women and their newborn babies reduced infection rates in the babies by two thirds. The trial was randomized and controlled; half the pregnant women agreeing to enter the trial were randomly selected to get AZT and the other half received no antiviral treatment. The incidence of HIV infection in their babies was then compared. AZT treatment caused such a dramatic reduction in infection that the trial was stopped early and after that AZT became part of the standard care for HIV-positive pregnant women. But in Asia and sub-Saharan Africa, where there will soon be an estimated six million

HIV-positive pregnant women, this treatment is too expensive to use routinely.

Following the American/French trial at least 18 similar trials were planned involving around 17 000 HIV-positive pregnant women. Fifteen of these trials (nine funded by the USA) were to be in developing countries where patients in the control groups would not be provided with any antiviral drugs. In contrast, in the two trials to be conducted in the USA, all patients had unrestricted access to antivirals. Only one American-funded trial planned in Thailand gave all the participating women antivirals and compared the effect of shorter (and therefore cheaper) courses of AZT with a regimen similar to that used in the original American/French trial.

This issue was highlighted in an article in the *New England Journal of Medicine* headed:

'Unethical trials of interventions to reduce perinatal transmission of the human immunodeficiency virus in developing countries'.

The article stated that the 15 trials 'clearly violate recent guidelines designed specifically to address ethical issues pertaining to studies in developing countries'.[3] According to these guidelines, 'the ethical standards applied should be no less exacting than they would be in the case of research carried out in (the sponsoring) country'.

This article raised a storm of protest from American and European veterans of tropical medicine, as well as several African doctors, who put the alternative view. They argue that any new drug regimen designed in the West has to be tested out in developing countries in conventional controlled trials. Results, they say, cannot simply be extrapolated from Western studies because of differences in lifestyle which might affect the outcome. In this case, for instance, developing countries afford fewer women access to antenatal care, particularly early in pregnancy when the treatment starts; intravenous infusion of the drug during labour may be impossible; and breast feeding, a known route of HIV transmission, is almost universal. Furthermore, they argue that in a developing country, where virtually no one has access to antivirals, it is ethical to carry out placebo-controlled trials, where the control group receive a harmless but useless drug, since no one is worse off than they would otherwise have been.

This controversy clearly has wider implications than the HIV field and will rage for a time to come. But already it has led to a rethink and to the careful planning of 'equivalency studies' where, as in the Thailand study, two drug regimens are compared with each other rather than with no treatment at all.

Turning the tables

Until the eighteenth century, smallpox had it all its own way. This virus caused the most lethal of all the acute childhood illnesses and epidemics swept through communities in regular cycles killing around a third of those it infected. But in 1715 the virus infected the beautiful young Lady Mary Wortley Montagu (Fig. 6.1), and although she survived, it left her face pock-marked and her eyes devoid of lashes. This turn of events gave her a keen interest in smallpox and led, a few years later, to the first successful prevention of the disease in Europe.[4]

Lady Mary's husband, Edward, was appointed British Ambassador to Turkey in 1716 and she travelled with him to Constantinople. Here she saw smallpox prevention in practice and wrote enthusiastically to her friend, Sarah Chiswell:

The smallpox, so fatal, and so general among us, is here entirely harmless by the invention of *engrafting*, which is the term they give

Fig. 6.1 Lady Mary Wortley Montagu (1689–1762). (Courtesy of Wellcome Institute Library, London.)

it. A set of old women perform the operation by scratching open a vein in the patient and putting into it as much of the smallpox venom as could lie on the head of a needle. On the eighth day the patient ran a fever which kept him in bed two or three days, and afterwards seldom showed pockmarks.[4]

She was referring to the practice of 'variolation' or 'inoculation', which probably originated in China. Healthy people inoculated with scrapings from smallpox pocks usually suffered a mild infection followed by life-long immunity. Lady Mary was so convinced that this was a safe and effective way to prevent smallpox that in 1717 she had her six-year-old son inoculated. Seven days later he developed a fever and about a hundred pocks. But the spots crusted and fell off leaving no scar; the experiment was a complete success.

When, in 1721, shortly after her return to England, a severe smallpox epidemic raged, Lady Mary used her position in high society to encourage the use of inoculation. But, as the following quote from one of her letters shows, she clearly had little faith in the medical profession helping her in this enterprise because of the income they stood to gain from a smallpox epidemic:

> I am patriotic enough to take pains to bring this useful invention
> into fashion in England, and I should not fail to write to some of our
> doctors very particularly about it, if I knew any one of them that I
> thought had virtue enough to destroy such a considerable branch of
> their revenue, for the good of mankind. But that distemper is too
> beneficial to them not to expose all their resentment the hardy wight
> that should undertake to put an end to it. Perhaps, if I live to return,
> I may, however, have courage to war with them.[4]

In 1721, Lady Mary's daughter was successfully inoculated in the presence of several influential doctors and so the practice became more generally known. Caroline, Princess of Wales, was interested but cautious. Before committing her two daughters to this new safeguard against smallpox she had it tested on six condemned prisoners in Newgate jail. They volunteered for inoculation on the understanding that if all went well they would be released and so it turned out—although the inoculation caused no symptoms at all in two of the prisoners, probably because they were already immune. But the Princess was still not satisfied, so she had 12 orphans from the Parish of St James inoculated at her own expense, before eventually allowing her daughters, the Princesses Amelia and Caroline, to be treated in 1723. This was successful and afterwards inoculation was widely practised in the UK and the USA. Between 1721 and 1728, 897 people were inoculated in Britain and 17 died of the effects; during the same period over 18 000 died of smallpox. So, if all those surviving inoculation were immune to smallpox, it was probably the safer option.

However, inoculation was obviously not entirely safe and was not universally accepted. Some doctors saw it as the wanton spreading of infection and, since at the time disease was widely believed to be a punishment from God, many of the clergy thought that it interfered with God's will. Despite this, it continued to be popular until 1798 when Edward Jenner published the details of a safer alternative.

Cowpox to the rescue

Edward Jenner (1749–1823) was a country surgeon–apothecary who lived and worked in Berkeley, Gloucestershire. In this rural setting Jenner became interested in cowpox which infected cows' udders and sometimes spread to the hands of dairy workers, causing localized blisters and a mild fever.

In his routine daily medical practice Jenner performed many inoculations and he noticed that, when inoculated, dairy workers often produced no response. In fact, whether or not they had suffered from smallpox in the past, they behaved as if they were already immune. Jenner deduced that cowpox infection must protect dairy workers from smallpox and he decided to test this out in a remarkably direct and practical way. In 1796 he inoculated eight-year-old

Fig. 6.2 Bronze statue by Guilio Monteverde showing Edward Jenner vaccinating a small boy. (Courtesy of the Wellcome Institute, London.)

James Phipps with live cowpox taken from the hand of a local milkmaid called Sarah Nelmes and then some weeks later tested the child's immunity by inoculating him with live smallpox (Figs 6.2 and 6.3). Fortunately, the child remained healthy showing that he was indeed immune to smallpox. In Jenner's own words:

> During the investigation of the casual Cow Pox, I was struck with the idea that it might be practicable to propagate the disease by inoculation, after the manner of the Small Pox, and finally from one

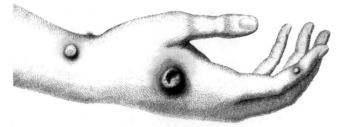

Fig. 6.3 Hand of Sarah Nelmes showing cowpox pocks. (Courtesy of the Wellcome Institute, London.)

human being to another. I anxiously waited some time for an opportunity of putting this theory to the test. At length the period arrived. The first experiment was made upon a young lad of the name of Phipps, in whose arm a little Vaccine Virus was inserted, taken from the hand of a young woman who had been accidentally infected by a cow. Notwithstanding the resemblance which the pustule, thus excited on the boy's arm, bore to variolous inoculation, yet as the indisposition attending it was barely perceptible, I could scarcely persuade myself the patient was secure from the Small Pox. However, on his being inoculated some months afterwards, it proved that he was secure.[5]

This historic human experiment, which was based on astute practical observation combined with intelligent and logical deduction, cut across all ethical and theoretical barriers to produce one of the most remarkable and important medical achievements ever recorded.

At first, some members of the medical profession argued against vaccination, maybe for the pecuniary reasons alluded to earlier by Lady Mary in her letter to Sarah Chiswell, but, thanks perhaps to Lady Mary's pioneering work which had set the scene, its obvious benefits were soon appreciated. Mass vaccination programmes were organized in Britain and Jenner's fame spread. In 1805 Napoleon gave the order for all his forces to be vaccinated and, in 1806, President Thomas Jefferson wrote to congratulate Jenner:

> Yours is the comfortable reflection that mankind can never forget that you have lived. Future nations will know by history only that the loathsome smallpox-pox has existed and by you has been extirpated.[6]

Neither Lady Mary nor Jenner knew why or how their particular type of smallpox prevention worked, but both rely on producing enough infection to stimulate an immune response without causing severe disease. Inoculation succeeded because a small dose of virus was administered by an unnatural route. Smallpox virus inoculated through the skin, rather than by the natural airborne route straight into the lungs, gets off to a slow start. By the time it spreads to a systemic infection, antibodies and immune T cells are ready to control and eliminate it. Material known as 'scabs' was collected from the actual pocks of active smallpox cases and used directly in inoculation. Even in the most careful hands this was a risky procedure as scabs were certainly capable of causing full-blown smallpox. (As we saw in Chapter 1, in 1763, during the fight for new territories in America, blankets deliberately contaminated with 'scabs' were distributed to Amercan Indian tribes with devastating effect.[7])

The success of Jenner's cowpox vaccination in protecting against smallpox relies on two facts. Firstly, cowpox and smallpox viruses are very similar—their genetic material is 95 per cent identical—and this is enough to trick the human immune system into thinking that they are one and the same. Secondly, cowpox virus only causes a mild localized disease in humans, and in doing so induces life-long immunity to smallpox as well as to itself. Smallpox vaccination was the first example of an attenuated (weakened) strain of virus used to protect against its more dangerous relative and it predated the now common use of this strategy by over a hundred years.

The original cowpox virus used by Jenner came from the hand of a milkmaid with the infection, but after this, direct arm-to-arm inoculation was used to pass the infection from one person to another. Later, another strain of pox virus—vaccinia—was introduced because it could be grown on the skin of calves, harvested, and used for several vaccinations. This vaccinia is the parent of the virus which is still used for vaccination, but its origins and the identity of its natural host have been lost in the mists of time.

Although vaccination soon became widespread in the Western world, where it had a dramatic effect in preventing smallpox, this was not the case everywhere. In non-industrialized countries, difficulties in organizing vaccine programmes, combined with the instability of the vaccine at high temperatures, meant that smallpox remained a significant killer on a worldwide scale until the WHO began its smallpox eradication programme in 1966.

The Franco-Prussian germ war

For several decades after the identification of microbes as the cause of infections, a battle raged over how best to harness the body's immunity against them. The two main protagonists were Louis Pasteur in Paris and Robert Koch in Berlin, whose rivalry reflected that of their respective countries at the time. The Germans tried to purify the antibacterial properties of serum, whereas Pasteur and his followers recognized the importance of white blood cells in the destruction of bacteria. Pasteur pioneered what later became the traditional approach for making viral vaccines using whole viruses which induce an immune response without causing disease.

Vaccine viruses may either be alive but weakened (attenuated) by prolonged growth under unfavourable conditions, or killed and therefore completely non-infectious. Armed with these we are now so successful in preventing infection that people born in the West after the mid-1970s are unlikely to have had any of the traditional childhood illnesses—measles, mumps, German measles—and the fear of paralytic polio is just a memory.

Rabies vaccine was one of the first to be produced. Like Jenner, Louis Pasteur made this without having any clear idea of the microbe he was dealing with. He used spinal cords from rabies-infected rabbits and inactivated the virus by leaving the material to dry out in a desic-cator. After 14 days all the infectious virus was destroyed and he found that he could render dogs resistant to rabies by injecting them on 14 consecutive days with rabbit spinal cord preparations dried for increasingly shorter periods. On the fourteenth day, the dogs were challenged with fully active virus and did not develop rabies.

In 1885, when Pasteur reckoned he was about two years away from having a vaccine against rabies safe enough for human use, he was persuaded to use his preparation on a nine-year-old boy from Alsace called Joseph Meister, who had been severely bitten by a rabid dog two days previously. There was no other hope for the boy, so Pasteur agreed to give him the same course of injections as he had given to dogs, with the last dose being the fully virulent virus. Like the dogs, the boy survived the injections and did not develop rabies. This success made Pasteur's vaccine famous and people bitten by rabid animals came to him for treatment from all over Europe.

Safe or sorry

There is little doubt that immunization has had an overwhelming effect on human health worldwide. It has reduced disease burden in every country in the world enormously and was the key to eliminating wild smallpox from the entire planet, and several other virus infections from the Western hemisphere. But vaccines are not entirely safe and so this immense achievement brings its own problems.

As childhood infections in the West recede into memory, vaccines become victims of their own success. With less and less chance of catching a virus, the small chance of one of the rare complications of its vaccine occurring becomes less acceptable. Thus for a single individual, logically it may be safer not to be immunized; certainly this was the case for smallpox, even before 1980 when the virus was officially declared eradicated. But if vaccination rates for any other virus drop in the West, then a significant non-immune population will emerge and epidemics will return. So, there is a conflict of interests between protection at a personal and at a population level. Each individual is dependent on others to maintain a high corporate level of immunity; those who choose not to be immunized are, in a way, parasites on the rest.

Public health services rely on mass vaccination to keep the nation healthy and consequently vaccine safety is paramount. Over recent years there have been many scares about vaccines—some real, others just rumours—but all potentially dangerous if they discourage parents from bringing their children for vaccination. Vaccines available before the early 1900s, like rabies vaccine, sometimes caused severe allergic reactions because they were made in animals and were difficult to purify and standardize. Later, vaccine viruses were grown in hens' eggs and this improved things slightly, but they were not entirely safe because they still contained animal proteins. These problems were largely overcome when cell culture was introduced for large-scale virus production—but this also brought other concerns.

From 1955 to 1963, polio virus and adenoviruses for vaccines were grown in cultured rhesus monkey kidney cells. In the 1960s, a new simian virus, called SV40, was isolated from monkey kidney cells; it was quickly found to contaminate these vaccines. Although chemical treatment with formaldehyde effectively inactivated the polio virus and adenoviruses in the vaccines, it did not always kill SV40. Tests showed live SV40 contaminating up to 30 per cent of vaccine batches and people given these vaccines tested positive for antibodies to SV40, indicating that they had been infected.

Researchers working on SV40 then found that it caused tumours in animals, so millions of people around the world had inadvertently been infected with a potential tumour-forming virus—98 million were given contaminated vaccine in the USA alone. Groups of these immunized people were followed to see if they developed more tumours than unimmunized people and thankfully they have not. But the problem resurfaced in 1992 when pieces of SV40 DNA were discovered in human tumours of bone, lung, and spinal cord—the very tumours that SV40 induces in infected animals. However, this work is controversial and the implications are far from clear. Many of the tumours containing SV40 did not come from people who received contaminated vaccine and other workers are unable to find any SV40 DNA in the tumours. Whatever the final outcome, the realization that cultured animal cells can harbour potentially dangerous viruses has led to human cells being used for vaccine production wherever possible.

Live attenuated vaccines usually induce more long-lasting immunity than killed preparations since they actually grow in the body, albeit briefly. But because of this, they have their own particular problems. They cannot be used during pregnancy for fear of infecting the developing baby and if given to patients with an immune deficiency, they may spread and establish a generalized or persistent infection. Also, reversion from attenuated to a virulent form of virus can occasionally occur. This is true of live attenuated polio virus vaccine which replaced killed polio vaccine in the 1970s. It has the great advantage of being taken by mouth rather than by injection and is now given to young children in three separate doses—but could there be problems? The vaccine virus infects cells lining the gut causing no illness but inducing an immune response. For a time after vaccination virus is excreted in faeces in large amounts and this probably spreads to other family members, so boosting their immunity too.

Each batch of attenuated polio vaccine is tested in animals to make sure that it has lost its capacity to cause paralytic polio before it is released for human use. But although it is extremely safe, it does cause paralytic polio in around one in every two million of those vaccinated. Somewhat surprisingly, comparison of genetic material from the paralytic and attenuated vaccine strains of polio virus shows very few differences—in most instances just two mutations. In cases of vaccine-associated paralytic polio, further mutations have returned these single changes back to their original form in the virulent virus.

In 1983, Philip Minor from the UK's National Institute for Biological Standards and Controls, set out to see how common back mutations of the vaccine strain of polio virus actually were.[8] He studied his own baby son David who was four months old when he had his first dose of oral polio vaccine. Minor collected all his baby's faeces and detected polio virus in them for the following 73 days. But it was molecular analysis of these viruses which really surprised him. He detected back mutations which increased the virulence of the virus as early as two days after immunization and more followed. To show that this was not just a fluke result, Minor repeated the same experiment two years later on his daughter, Elizabeth, and came up with the same answer. Up until then polio mutants had only been found in rare cases of vaccine-associated paralytic disease, but thankfully the Minor children remained healthy.[9]

We now know that back mutations occur in almost all those vaccinated, but why these viruses so rarely cause paralysis is not yet clear. However, with this level of reversion to a potential disease-causing virus and circulation of vaccine virus in the community complete eradication of paralytic polio and the virus will probably not be feasible using live attenuated vaccine. For this we may need to change back to the earlier killed preparation.

Recently, a scare over the use of the measles/mumps/rubella vaccine (MMR) in the UK reverberated around the world. It all started with a report published in the medical journal *The Lancet* suggesting an association between MMR (particularly the live measles component), chronic bowel disorders, and the behavioural disorder, autism.[10] The report received massive media coverage and this produced an immediate fall in MMR uptake. Despite the fact that scientific experts quickly showed the associations to be unfounded, the damage was done: 2000 fewer children than usual were immunized in the following three-month period. Measles is all but eliminated in the UK, but it is still a common and serious infection in some non-industrialized countries where the mortality rate can be up to 25 per cent. Non-immunized children in the UK will remain susceptible all their lives, so if they contract the disease while travelling abroad this could form a nucleus for a future measles epidemic among the non-immune in the UK.

Fruit, vegetables, and naked DNA

Not all viruses can be prevented by conventional vaccines. There are big killers out there like respiratory syncytial virus, rotavirus, and HIV which are still spreading fast. But now, with the revolution in recombinant DNA technology, there is a good chance that they too can be defeated. From their genetic sequences, immunologists can identify the particular bits of their proteins (peptides) which are targets for the immune response. These essential components are then made in the laboratory and used as new generation 'subunit' vaccines. These are purer and safer than classical vaccines because they do not contain whole viruses or any material derived from them.

The first new generation vaccine to come on line was against hepatitis B virus. Yeast cells, genetically engineered to contain the gene for hepatitis B surface protein, grow easily in culture and produce large quantities of the pure protein—certainly a much safer option than the original vaccine which was purified from the blood of hepatitis B virus carriers.

Other modern vaccines rely on gene swapping. Here the gene for the essential protein is cloned into a harmless carrier virus. When inoculated, this virus infects cells and produces the foreign vaccine protein along with its own so stimulating an immune response. Already large viruses like pox and herpes viruses have been rendered harmless by removing vital genes for their growth and spread and sticking HIV genes in their place. Now even 'naked DNA' (that is, the gene without a carrier virus), when injected straight into muscle, has proved effective in test animals. If these techniques work in humans, soon we may have a one-shot vaccine—be it a string of peptides, a carrier virus containing key genes from other microbes, or simply naked DNA—which renders us immune to a whole range of microbes.

For viruses like rotavirus, which infect the gut, it is best to deliver the vaccine straight to the natural site of infection where it can induce local immunity. This means that these vaccines, like polio, should be ingested rather than injected. Since this would obviously be more preferable to children, and is safer than injection, research on oral vaccines is a growth area at the moment. Plants have been genetically engineered to express viral proteins and we already have transgenic bananas expressing hepatitis B surface protein and potatoes expressing rotavirus proteins. So the concept of an edible vaccine is fast becoming a reality.

Prevention is better than cure

There is little doubt that the best and most cost-effective way to prevent HIV infection and spread on a worldwide scale is to produce a vaccine. But this has proved more difficult than expected and has become one of the biggest challenges to scientists today. In 1997, President Clinton pledged an AIDS vaccine within 10 years and set up a new Vaccine Research Center to spearhead the effort.[11] However, scientists are divided on how best to proceed. On one hand, there are those who believe in a practical trial-and-error approach. They argue that since HIV infection is fatal, and people with it are dying every day, any strategy with a chance of success should be tested in clinical trials. The other camp believe that until more is known about protective immunity against HIV, and which viral genes to target, there is little point in undertaking clinical trials. They say that trials which are not backed up by solid scientific evidence are likely to fail and may even be dangerous. Furthermore, they would waste the limited funds available for research and undermine public confidence.

The simian immunodeficiency virus (SIV) which naturally infects Rhesus monkeys, is harmless in its natural host, but infection of related macaque monkeys causes simian acquired immunodeficiency (SAIDS). This is very similar to AIDS in humans and is the best animal model we have for testing out vaccine strategies. The traditional approach certainly works in macaques where SIV, which has been attenuated by removing genes which are responsible for maintaining a high viral growth rate, is the only vaccine tested so far which completely protects monkeys from challenge with fully virulent virus. The next logical step is to test out a similarly attenuated HIV vaccine in humans, but there are clear safety concerns. The attenuated virus might still cause AIDS, but after a longer latent period, or it might mutate or recombine with the naturally-occurring virus and revert to the fully virulent form. Most doctors, scientists, and HIV-infected people think that this is too risky to try out. But in 1997, a call for volunteers for just such a trial from the International Association of Physicians in AIDS Care, appeared on the internet. The trial is being masterminded by Charles Farthing from Los Angeles, California, who is said to be 'progressively irritated by the lack of movement towards clinical trials of an attenuated HIV vaccine'. He already has nearly 300 volunteer health workers prepared to be vaccinated, and he plans to begin the trial in the near future.[12]

At the moment, scientists are busy trying to identify viral proteins which are targets for protective immunity in vaccinated macaques. If this can be unravelled, then a subunit vaccine containing individual viral genes rather than whole virus could be tried. However this approach is proving difficult. The receptor molecule on the surface of HIV, called gp120, which binds to CD4 cells, is essential for infection, and antibodies against gp120 prevent HIV infection of cells in tissue culture. Logically therefore, a gp120 vaccine which induces antibodies should prevent infection in animals. Indeed in some experiments, chimps with antibody against gp120 are protected from later challenge with HIV. Unfortunately, in other experiments, they are not and clinical trials in humans give disappointing results.

Twenty-two different gp120 vaccines have so far been tested in small trials involving a total of around 2000 volunteers likely to be exposed to HIV through high-risk behaviour. Although those vaccinated did develop antibodies against gp120, these were generally ineffectual in blocking infection in laboratory tests. A total of 19 of those vaccinated became infected with HIV—no different from numbers in the control unvaccinated group. Despite this failure, other large-scale trials using gp120 vaccines are under way in the West and in developing countries. No one is very hopeful that this will provide the answer to HIV prevention, but at least we may get some clues about how to proceed in the future.

What's the problem?

When HIV was first discovered in the 1980s, scientists were optimistic about producing a vaccine fairly quickly. After all, HIV only has a handful of genes, so it should be easy to work out which are targets for the immune response. However, HIV is a devious customer and many unforeseen problems arose. One is the extreme variability of the virus; its unique flip from RNA to DNA after infection of a cell is a highly error-prone step which generates lots of mutants to outflank the immune system's troops. So not only are viruses in one person always changing but viruses collected from different people are quite diverse. This makes it difficult, if not impossible, to design a single vaccine which will protect everyone from infection.

Another difficulty is the persistence of the virus. Most successful vaccines are aimed at viruses which cause acute infections, where the vaccine induces immunity to future disease. These viruses probably still infect immunized people from time to time but after immunization, infection is controlled at an early stage and the virus eliminated without causing disease. Since the first step in HIV's strategy is to insert its genetic material into the DNA of host cells, thereafter it is probably not possible for any immune response to remove it. So the only approach is to prevent HIV infection entirely and if antibodies against gp120 cannot achieve this then it is difficult to see what will.

To follow this line of attack it is vital to know more about how the virus infects naturally—not just that it is transmitted in genital secretions, but how it gets a foothold in a new host. We know that the virus grows rapidly and spreads throughout the body before the immune system is alerted to its presence. It is now clear that dendritic cells, which live in skin and mucous membranes, are at least partly responsible for this. As their name suggests, these cells have long spidery processes, and their job is to trap foreign material such as invading viruses which penetrate body surfaces. Then, they carry the invaders to the local lymph glands where they show their catch to resident CD4 T cells. This vital interaction kick-starts the immune response. But HIV can actually infect and grow in the dendritic cell which traps it and then it is ready and waiting to infect each CD4 T cell which the dendritic cell interacts with.

It is not easy to study the very beginning of HIV infection in humans because only rarely do we know when someone first becomes infected. Another approach is to compare experimental SIV infection in immunized and non-immunized macaque monkeys to find out exactly how the attenuated virus protects. One suggestion is that infection of dendritic cells with attenuated virus prevents their subsequent infection with virulent challenge virus. If this is the case, then the vaccine strain of virus would have to stay in the body to maintain protection long term. This carries with it the ever-present danger of it causing AIDS at some time in the future.

Understanding the precise natural immune responses which correlate with protection is also essential for producing a human vaccine which does the same job. Here, two special groups of people are giving us valuable information. Andrew McMichael and his team from Oxford University were intrigued to find small groups of commercial sex workers in the Gambia and Kenya who consistently tested HIV negative despite astronomically high levels of infection in their colleagues and clients.[13] Condom use was ruled out as the reason for this protection because the HIV-negative women had just as many STDs as HIV-positive sex workers. Also, their cells could readily be infected with HIV in the laboratory, so genetic resistance to the virus was not the explanation. The team found that although these women had no HIV antibody, they did have killer T cells targeted to HIV infected cells. But to generate killer T cells they must have met HIV; somehow their immune systems had conquered and routed the virus. Perhaps frequent exposure to very low doses of virus had built up enough immunity to prevent HIV getting that vital early foothold. Whatever the mechanism, McMichael now aims to mimic this killer T cell response with a combined vaccine (a shot of naked DNA, followed by a boost with HIV genes in a canary pox virus), in the hope of inducing complete protection.

Another fascinating group are the few people still alive and well 20 years or more after they caught HIV. A case in point is Bob Massey, an Episcopalian minister and haemophiliac who was infected from contaminated factor VIII in 1978, the year he graduated from college. He has been the subject of a TV documentary entitled 'Surviving AIDS'[14] which highlighted the complex medical and moral issues surrounding the subject. Massey is one of a small band (about 5 per cent) of HIV-positive people, called long-term non-progressors, who break all the rules. He has undetectable viral load and a normal CD4 count, although he is infected and has HIV antibody. Like the African commercial sex workers, doctors find he has HIV-targeted killer T cells—enormous numbers of them in fact—which must be forestalling the virus' every attempt to get the upper hand. His blood, studied intensively by Bruce Walker and his team at the Massachusetts General Hospital, Boston, USA, is helping scientists to understand what is required to control this devious virus. No one can tell Massey whether he will stay healthy or one day come down with AIDS, but for now he lives a normal life, and the TV programme ended with him christening his healthy HIV-uninfected baby daughter.

Self help

While high-powered science and expensive drugs are brought to bear on the AIDS problem in the West, in Africa people are setting about helping themselves as best they can. In Uganda, where the AIDS epidemic has been devastating and only $3 per person a year is spent on health care, there is little hope of any treatment. The emphasis is on prevention and educational programmes set up in 1989, when the country was at the forefront of the epidemic, are now bearing fruit. Schoolchildren learn of the risks and, in some areas at least, this has at last led to a fall in levels of HIV infection among young men and women.

Another approach is exemplified by Noerine Kaleeba who set up The AIDS Support Organisation (TASO) after her husband died of AIDS in 1987. Starting with one small self-help group, this organization now has 17 clinics throughout Uganda operating under the slogan 'living positively with AIDS'. The clinic in Entebbe is cramped and dilapidated but staff radiate friendliness and optimism. It provides free counselling and care for 1200 people with AIDS, most of whom are women. Encouragement not to regard their infection as a guilty secret has generated a refreshing openness about AIDS which is rare in Africa. This kind of enterprise, which is mirrored by other small groups all over the world, supports and comforts people with AIDS and their families and is an essential stopgap until scientists eventually come up with a method of prevention or cure which is affordable for all.

The beginning of the end

The goal of the WHO is to provide 'a healthier life style for all' and part of this is to rid the world of deadly microbes. Smallpox was a good place to start because it killed around one third of those it infected and although it was eradicated from the UK in the 1930s and from the the USA and most of Europe by the 1940s, imported cases were a constant threat, so strict surveillance and quarantine laws had to be kept in force. The WHO worldwide smallpox eradication campaign began in 1966 at a time when the disease was endemic in 31 countries and in that year 131 776 cases were reported, with 20 per cent fatalities. Given that this probably underestimated the total number of cases by a hundredfold, smallpox was obviously still a major world health problem and governments, rich and poor, were united behind the campaign.

Operationally, the WHO team, headed by Don (D.A.) Henderson, defined a smallpox endemic area as anywhere with over 5 cases per 100 000 population and less than 80 per cent of people vaccinated. They assigned each of these areas to one of four isolated endemic zones with little chance of cross infection between them— Brazil, Indonesia, sub-Saharan Africa, and the Indian subcontinent (Fig. 6.4). They aimed to interrupt the spread of the virus since, because smallpox is an acute infection with no animal reservoir, if they could break the chain of infection the virus could not survive. Priorities were to improve vaccination rates and increase case reporting, to trace all contacts of cases, and to isolate infectious people. This required a massive investment; fieldworkers and laboratory staff worked hand in hand to produce an effective, heat-stable vaccine, to vaccinate remote and sometimes reluctant people, and to diagnose acute cases and identify contacts.

The campaign was so effective that by 1973 only six endemic countries remained—India, Bangladesh, Pakistan, Nepal, Botswana, and Ethiopia. By 1975 Ethiopia was the only endemic area and by 1976 infection was finally cleared. Before any country was declared smallpox-free a 'quarantine' period had to elapse during which surveillance continued unabated. Monkeypox and chickenpox often look remarkably like mild smallpox, so all

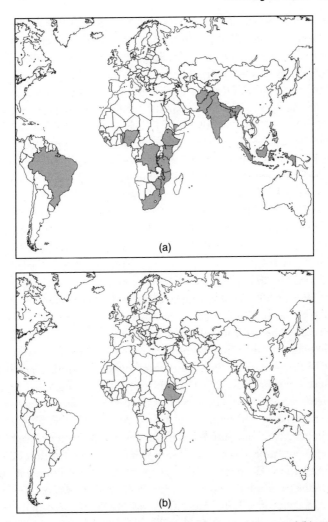

Fig. 6.4 World maps showing smallpox endemic areas in (a) 1968 and (b) 1975.

rumoured cases had to be checked. The following figures from India, for a single month in 1976, show the staggering extent of the operation: 668 332 villages, including 3 051 753 households, were searched by 115 347 workers, who were themselves overseen by 29 046 supervisors. The work was monitored by 8048 assessors who found that 97 per cent of the 107 409 villages they assessed had, in fact, been searched. All this effort uncovered 41 485 cases of chickenpox but no smallpox. Thus India was declared smallpox-free.

Unbelievably, the last two recorded cases of smallpox in the world were in Birmingham, UK. In 1978, a photographer who worked in the Department of Anatomy at the University of Birmingham Medical School died of the disease; one other person caught the infection from her but recovered. Neither the photographer nor anyone else in the Anatomy Department worked with smallpox virus, but large quantities were grown in the Department of Microbiology situated in the same building on the floor below. An enquiry into the catastrophe found that the photographer was infected with the strain of virus used in the smallpox laboratory, where safety procedures were reported to be 'far from satisfactory'.[15] The virus probably travelled through an air duct leading from the smallpox preparation area to the floor above where the enquiry team found an ill-fitting air duct inspection panel close to a telephone which the photographer used several times a day. The report pointed the finger at smallpox researchers for breaking the safety rules, the Microbiology Department for not enforcing them, and the inspector from the Dangerous Pathogens Advisory Group for failing to realize the extent of the hazard. The head of the department at fault killed himself, so bringing the death toll from this regrettable incident to two.

Worldwide smallpox eradication was finally declared in 1980. The whole enterprise cost $312 million, of which an estimated $200

million would otherwise have been spent on routine control programmes. There are now supposedly only two places in the world
where smallpox virus remains—in safe freezers at the US Center for
Disease Control and Prevention, Atlanta, and at the Research
Institute for Viral Preparations in Moscow. Now scientists argue
about whether these last stocks should be retained or destroyed.
Those who vote for keeping the virus argue that studying it may
help in understanding other pox virus infections like monkeypox
which is now becoming more of a problem. Also, the rising threat
posed by bioterrorism means that it may be necessary to develop
new vaccines sometime in the future. Others use the conservation
argument, suggesting that deliberate destruction of any living
organism could set an unfortunate precedent for the future.
However, those wishing to destroy the virus maintain that any store
of smallpox is unsafe and the Birmingham incident is a case in
point. Besides, the entire DNA of several pox viruses are cloned and
this would seem to counter any reason for keeping infectious particles. The WHO decided that the last remaining smallpox viruses
would be destroyed by the end of the century, but meanwhile the
US Institute of Medicine had other ideas. They suggest that stocks
should be retained for further vaccine work—so the virus was
granted a last-minute stay of execution.

The incredible success of the smallpox eradication programme
showed, for the first time, that man could triumph over a lethal
virus and this heralded a new era in the fight against infectious diseases. The WHO now has a hit list—polio and measles are well on
the way to extinction, while the campaigns against rabies and
hepatitis B are at an earlier stage.

When it comes to elimination, each virus has its own particular
defences to be overcome; some lurk in animal hosts, some set up

persistent infections which will be difficult to ferret out, and for others, we do not yet have a foolproof vaccine. However, the long-term benefits of eradication, in financial as well as health terms are clear to see and so although success may be several decades away it is certainly the way forward.

The other line of attack we have is to develop antiviral drugs which kill off established infections and here again there is optimism. We are likely to see an explosion of these products over the next 10–20 years. Hand in hand, these two approaches will play a major role in achieving the WHO goal to provide a healthier lifestyle for all.

Conclusion: The future— friend or foe?

Here we are at the beginning of the twenty-first century and in the midst of a technological revolution, but viruses still flourish. They evolve rapidly, exploiting every opportunity to promote their own survival and leave us, with our much slower rate of change, lagging far behind. Is this always going to be the case? Will we ever conquer viruses? Or, on the other hand, will new lethal viruses continue to emerge? Could a new virus eliminate the human race?

Broadly speaking, in previous chapters we have looked at three types of viruses which threaten our health. In evolutionary terms these are old, young, and newborn viruses. Now, in this last section, we can assess each for its threat to our future existence.

Old and subdued

Ancient viruses, like the herpes family, inhabited our evolutionary predecessors from the dawn of civilization and were passed on when mammals evolved some 80 million years ago. Back then, they were probably aggressive beasts, but over time they have co-evolved with us to reach a state of mutual tolerance. Now almost the whole human population is infected without being aware of it. Once on board, these viruses bed down for a lifetime of cohabitation. Infection is not generally life-threatening in a healthy host, but that is not to say it is always completely harmless. Some, like herpes simplex reactivate occasionally to give unpleasant symptoms while others, such as Epstein–Barr virus, can cause tumours. And when released from the tight grip of the immune system, either by HIV infection or immune suppressing drugs, they can kill.

Ancient viruses are not a serious threat to the human race, but are they of any use? The silent, persistent type of infection calls to mind viruses like Hanta and Lassa fever viruses. They may cause devastating infection in humans, but in their primary hosts (the deer mouse for Hanta, the African brown rat for Lassa fever), they are persistent and symptom free. Frank Ryan, in his best-selling book, *Virus X*, refers to this scenario as 'aggressive symbiosis'.[1] He disputes the fact that persistent viruses are merely parasites and instead suggests that they play a useful role in protecting their primary hosts. He argues that the sophisticated immune system in mammals could have got rid of persistent viruses long ago if there was no advantage to be gained from keeping them. So mammalian immunity must have evolved to control but not eliminate these ancient viruses which are somehow used in self-defence.

As an analogy, Ryan describes ants nesting in acacia trees on savannahs surrounding African rain forests. The tree provides the ants with food in the form of nectar and its long sharp spines protect them from predators. In return, when a browsing animal like a giraffe comes along intent on eating the tree, ants swarm out and bite its soft mouth parts, so persuading the animal to move on to another tree for its meal. This action may kill thousands of ants but in the long run preserves their species. In the world of microbes, the ant colony represents a persistent virus safely lodging in a host (the tree). The virus plays its part in protecting the host by infecting and causing disease in other species which threaten the host's existence.

It is difficult to see how such a relationship could evolve since in the battle to control viruses the host's immune system is generally in response mode. That is to say, since viruses change rapidly and we cannot second guess these changes, we are left with no option but to respond. In fact, the sophistication of our immune system is entirely due to this adaption process recurring over and over again during evolution. In this cat and mouse game, a unilateral decision by the host not to respond would not be an option. Much more likely, viruses have evolved ways of evading host immune control. However, having said that, there certainly are examples in the animal kingdom which fit the aggressive symbiosis model.

Hanta and Lassa fever viruses appear to protect their animal hosts by infecting man when he comes too close for comfort, thereby perhaps discouraging him from invading their territory. No very convincing human examples come to mind where we are protected by infecting other species with our persistent viruses. But perhaps man is so successful that any animal species susceptible to human viruses are already extinct. Or maybe our persistent viruses will turn out to be our best protection against invading extraterrestrials! Whatever the reason for their persistence, these viruses are unlikely to be a threat to our continued existence and may even prove beneficial.

Young and lively

'Young' viruses cause epidemics and pandemics. The cyclical pattern which characterizes their infection began when our ancestors settled into fixed communities based on agriculture some 10 000 years ago. Viruses took advantage of the close proximity between man and his farm animals and jumped species to infect a new virgin population. Domestication of cattle probably offered rindepest virus the opportunity to infect humans causing what we now recognize as measles and smallpox probably originated from cowpox virus. For survival, these viruses require a regular supply of around 7000 susceptible people, which is equivalent to a town with half a million inhabitants. So once urban living became the norm their success was ensured.

Acute virus infections like measles have become steadily milder over the intervening 10 000 years but still, until recently, they clocked up a massive death toll. Continuing co-evolution would probably render these viruses less of a threat, but fortunately we do not have to wait that long. With the invention of vaccination in 1796, Jenner tipped the balance in the battle between man and virus in our favour. Since then we have succeeded in controlling many viruses with vaccines, and using new and ingenious technologies we can now attack even the most recalcitrant of them. So the threat posed by epidemic virus infections should continue to recede, although there may be a new cloud on the horizon.

Recent problems caused by intensive farming practices parallel in animals what happened to our ancestors in their new farming communities. A cage of farmed salmon is as easy prey to a fish virus as a crowded town is to a human virus. Even more so in fact because farmed fish tend to be inbred and lack genetic variation; similarly, a battery of chickens, field of corn, or barn of pigs. If one is susceptible then most likely they all are and a virus can rip through and kill the lot.

There are over 14 million pigs in Holland packed into massive barns at a density of some 9000 animals per square kilometre. This is asking for trouble. The most recent epidemic of swine fever in 1997 killed over six million pigs; actually most of these died healthy—killed by farmers to prevent the virus spreading. The whole disastrous affair cost the Dutch Government 400 million pounds in compensation. An outbreak of swine foot and mouth disease, caused by an airborne virus, would be even more disastrous; it could eliminate most of Europe's pig farms in one go. Fortunately, the 2001 foot and mouth disease outbreak in the UK, although devastating to cattle and sheep farmers, was caused by a new strain which does not infect pigs efficiently.

As if this is not enough, these overcrowded farms are a hot bed of genetic mixing for flu viruses. A new strain capable of causing a human pandemic could emerge at any moment. The Dutch Government is trying to avoid a catastrophe by restricting the numbers of pigs on each farm and movement of animals between farms. This may help, but still in neighbouring Belgium and Germany, where pig farming is just as intense, no such restrictions exist and the threat continues.

Newborn and aggressive

New emerging viruses are now our major adversaries. They are unpredictable, cropping up at random in remote parts of the world with increasing frequency. Because we have no previous experience of them, the diseases they cause are severe and often fatal.

These viruses are not entirely new. Most, like HIV, crossed the species barrier from their natural host to infect man. Others such as flu mutate regularly to outwit our immune defences. So what is the likely influence of these viruses on man's destiny?

Wherever we have colonized we have disturbed natural ecosystems and caused long-lasting knock-on effects to surrounding plants, animals, and microbes. Viruses are quick to exploit these imbalances and the result all too often is new human disease. The effects are more pronounced in tropical than temperate climates because the huge diversity of animal and plant species supports a greater variety of microbes. So most 'new' infections originate in tropical areas and one of the newest, monkeypox, has somewhat sinister origins.

Smallpox has been eliminated from the wild and measles and polio are well on their way to extinction. These successes alone save around seven to eight million lives every year. It seems perverse to argue that the world is worse off without these killers, but some scientists say exactly that. Their logic is that some other microbe, perhaps worse, will fill the gap in the ecosystem vacated by an extinct virus. So better the devil you know—and have evolved resistance to. Now this fear may be proving true with smallpox.

The first known monkeypox case in humans occurred in 1970 in the Democratic Republic of the Congo, Central Africa, and in the following decade only 55 cases were reported. The virus is not carried by monkeys, as the name suggests, but by squirrels in the rain forests of West Africa, and once humans are infected from the primary host, person to person transmission can follow. Symptoms resemble smallpox with a pustular rash and high mortality. Smallpox vaccination protects against monkeypox, so during and after the smallpox eradication campaign of the 1970s, the vaccinated population was not at risk. There were only a handful of cases and nobody took them very seriously. However, smallpox vaccination is no longer routine and the situation has changed dramatically.

In 1996, there was an outbreak of monkeypox in the Congo with 511 suspected cases. The virus spread from patient to household contacts in a frighteningly high 42 per cent of cases. This had all the hallmarks of a serious emerging infection with the capacity to spread rapidly in a population where immunity to smallpox was waning. Although some suspected cases later turned out to be chickenpox, the WHO are now carrying out field studies to monitor the spread of this disease which could be picking up where smallpox left off. With this in mind perhaps we should think twice before eradicating anything but the most deadly viruses from the wild.

At the moment we are severely out of balance with the natural environment and this is directly responsible for the recent rise in 'new' virus infections. The situation will only change for the better if we redress the balance and restore harmony with our surroundings.

Ten thousand years ago a change in lifestyle caused new virus infections which we have struggled to control ever since. Our immune system has learnt to cope with these infections, which have subsequently become less severe. But it has been a slow process and during the adaption millions of lives—and mainly young lives—have been lost. Those who recovered had lifelong protection and so viruses constantly needed a fresh supply of susceptible (young) people in the population to maintain their cycles of infection. We adapted to this loss by increasing the size of our families to ensure that some offspring survived.

Over the last century, the introduction of public health measures, modern medicine, and vaccination have prevented many deaths, and now the balance is again disturbed and we have a population explosion. The world population has reached six billion and this figure is rising rapidly. Consequently we have to colonize new territories, be it the Australian outback, the pampas of South America, or the rain forests of Africa. And here again we meet new viruses and go down with new and dreadful diseases. So population control could be the key to breaking the cycle. Alongside this, continued elimination of deadly viruses and development of new antiviral agents should assure man's continued survival—unless of course there is an inadvertent or deliberate release of a killer virus.

'Designer' killer

For the complete elimination of the human race, an entirely new virus must appear which, like smallpox, is robust, highly infectious, spread by aerosol, and causes lethal disease. If a recent BBC documentary[2] is to be believed, somewhere in Russia just such a 'new' virus is being made. This is an amalgam of the deadliest parts of Ebola and smallpox—aptly named 'doomsday virus'. A virus with these qualities stands a chance. Now that the concept of the global village is a reality, even a virus with an incubation period of just a few days can board a plane in an apparently healthy carrier and reach any country in the world before revealing itself. But humans possibly still hold the trump card—genetic diversity.

Our genes are so mixed and varied that our response to infection is equally so. In any epidemic there are always those who get infected but suffer no, or only mild, disease. So a proportion, however small, may survive attack with even the most lethal virus.

Turning the tables

Viruses exploit us very successfully to promote their own survival, but can we in turn exploit them? Already scientists are pioneering clever ways of putting viruses to good use; a simple example is the recent discovery of a virus which naturally infects the corn ear-worm. This worm is a particularly troublesome pest in cornfields of the USA, and its close relatives, the legume pod borer and the cotton bollworm, devastate crops in Asia and Africa. Since the virus targets and destroys the gonads, infected worms metamorphose into sterile moths. Now investigations are under way to see if this rather drastic contraceptive device can be used for pest control.[3]

Being small and relatively simple bits of genetic material which are adept at gatecrashing their way into cells, viruses have taught us a great deal about the concept of gene therapy. Here, in theory, an abnormal or missing gene is replaced by a gene which functions normally. As soon as the gene responsible for a disease has been cloned, normal copies can easily be produced in the laboratory. But it is much more difficult to deliver them to the right type of cell in the body and this is where viruses can help out. After all, why not use a ready-made parasite with generations of experience in penetrating cells as long as it can be rendered harmless? So scientists in laboratories around the world are searching for the ideal virus vector to carry designer genes into cells.

In haemophilia, for example, where a defective gene for blood clotting factors VIII or IX causes a bleeding disorder in otherwise perfectly fit men, the benefit would be enormous. At the moment sufferers need regular injections of the missing factor, prepared from donor blood, to restore normal clotting. This treatment has to be repeated regularly and carries the risk of infection with a blood-borne virus. It would be much more satisfactory to restore the clotting factor production permanently by replacing the defective gene with a functioning one. This has been successful in dogs with haemophilia using a modified adenovirus vector.[4] But this strategy is at the experimental stage and there are still several problems to iron out—not least, the immune response which recognizes the virus vector as foreign and attempts to eliminate it.

As well as replacing genes which are defective, gene therapy, in the form of the 'magic bullet' approach, is being used to treat killer diseases like cancer. Most of these strategies are still under test in the laboratory, but one type of magic bullet is already being used in clinics for the treatment of brain tumours. Here a modified herpes simplex virus is not used to carry foreign genes into cells but to kill tumour cells directly, simply by growing in them.[5] The trick is that engineered virus only grows in cells which are actively growing and dividing themselves. Normal brain cells do not divide, but brain tumour cells do. So, when injected into tumours the virus grows in, and kills, cancer cells. It spreads until reaching the edges of the tumour where it meets normal brain cells; then it has to stop, sparing normal brain tissue. The great advantage of this system is that the brain is protected from immune mechanisms which function elsewhere in the body, so here therapeutic viruses stand a better chance of survival until their job is done.

Traditionally the bad guys which have spawned only disease and misery, viruses are now being trained to be useful. So perhaps in the twenty-first century we can look forward to a change in our view of viruses.

References

Introduction: The deadly parasite

1 *Outbreak,* a film by Warner Brothers (Time–Warner Entertainment Company).

Chapter 1: Bugs, germs, and microbes

1 Medawar, P.B. and Medawar, J.S. (1983). Aristotle to Zoos. *A philosophical dictionary of biology.* Harvard University Press, Cambridge, Massachusetts, USA.

2 *The Concise Oxford Dictionary* (9th edn) (ed. D. Thompson). (1995). Clarendon Press, Oxford.

3 Radford, T. (1996). Influence and power of the media. *Lancet,* **347**, 1533–5.

4 Cramb, A. (4 December 1996). Food poisoning claims sixth life. *The Daily Telegraph.*

5 *The Independent on Sunday Review.* (20 September 1998). Lab notes, p. 62.

6 Pavord, A. (1999). *The tulip.* Bloomsbury Publishing, London.

7 Professor Norman Noah, Department of Public Health and Epidemiology, King's College London, Bessemer Road, London, SE5 9PJ. *Personal communication.*

8   Bosch, X. (1998). Hepatitis C outbreak astounds Spain. *Lancet,* **351**, 1415.

9   McAlister Gregg, N. (1941). Congenital cataract following German measles in the mother. *Transactions of the Ophthalmological Society of Australia,* pp. 35–46.

10 McNeil, W.H. (1994). Plagues and peoples. Penguin Books, London.

Further reading

Collier, L. and Oxford, J. (1993). *Human virology.* Oxford University Press, Oxford, New York, Tokyo.

Ewald, P.W. (1994). *Evolution of infectious disease.* Oxford University Press, Oxford, New York.

Grafe, A. (1991). *A history of experimental virology.* Springer-Verlag, Berlin, Heidelberg.

Holland, P.V. (1996). Viral infections and the blood supply. *New England Journal of Medicine,* **334**, No. 26, 1734–5.

McGeoch, D.J., Cook, S., Dolan, A., Jamieson, F.E., and Telford, E.A.R. (1995). Molecular phylogeny and evolutionary timescale for the family of mammalian herpesviruses. *Journal of Molecular Biology,* **247**, 443–58.

Oldstone, M.B.A. (1998). *Viruses, plagues and history.* Oxford University Press, Oxford.

Poupard, J.A. and Miller, L.A. (1992). History of biological warfare: catapults to capsomeres. *Annals of New York Academy of Sciences,* **666**, 9–20.

Strauss, E.G., Strauss, J.H., and Levine, A.J. (1990). Virus evolution. In: *Virology* (2nd edn) (ed. B.N. Fields, D.M. Knipe, *et al.*) Raven Press Ltd, New York.

Waterson, A.P. and Wilkinson, L. (1978). *An introduction to the history of virology.* Cambridge University Press, Cambridge, London, New York, Melbourne.

Watson, J. and Crick, F. (1970). *The double helix.* Penguin Books, London.

Chapter 2: New viruses or old adversaries in new guises?

1   Nowa, R. (1995). Cause of fatal outbreak in horses and humans traced. *Science,* **268**, 32.

2   Murray, K., Selleck, P., Hooper, P., Hyatt, A., Gould, A., Gleeson, L., *et al.* (1995). A morbillivirus that caused fatal disease in horses and humans. *Science,* **268**, 94–7.

3   Wells, R.M., Estani, S.S., Yadon, Z.E., *et al.* (1997). An unusual Hantavirus outbreak in Southern Argentina: person-to-person transmission? *Emerging Infectious Diseases,* **3:2**, 171–4.

4   Ban, E. (1997). New virus leaves questions hanging. *Nature Medicine,* **3**, 5.

5   Lupton, D. (1994). *Moral threats and dangerous desires: AIDS in the news media.* Taylor and Francis, London.

6   (1981). Pneumocystis pneumonia—Los Angeles. The morbidity and mortality weekly report, **30**, No. 21.

7   (1981). Kaposi's sarcoma and pneumocystis pneumonia among homosexual men. The morbidity and mortality weekly report, **30**, No. 25.

8   Barré-Sinoussi, F., Chermann, J.C., Rey, F., *et al.* (1983). Isolation of a T-lymphotropic retrovirus from a patient at risk for acquired immune deficiency syndrome (AIDS). *Science,* **220**, 868–71.

9 Weiss, R.A. and Wrangham, R.W. (1999). From pan to pandemic. *Nature,* **397**, 385–6.

10 De Cock, K.M. (1996). The emergence of HIV/AIDS in Africa. *Revue d' Epidemilogie et de Santé Publique,* **44**, 511–18.

11 Grosskurth, H., Mosha, F., Todd, J., Mwijarubi, E., Klokke, A., Senkoro, K., *et al.* (1995). Impact of improved treatment of sexually transmitted diseases on HIV infection in rural Tanzania: randomised controlled trial. *The Lancet,* **346**, 530–36.

12 Clumeck, N., Taelman, H., Hermans, P., Piot, P., Schoumacher, M., and de Wit, S. (1989). A cluster of HIV infection among heterosexual people without apparent risk factors. *The New England Journal of Medicine,* **321**, No. 21, 1460–2.

13 Brettle, R.P. (1995). *Human immunodeficiency virus: the Edinburgh epidemic.* (Edinburgh University dissertation for the degree of MD.)

14 May, R.M. and Anderson, R.M. (1987). Transmission dynamics of HIV infection. *Nature,* **326**, 137–42.

15 Gajdusek, D.C. (1957). Degenerative disease of the central nervous system in New Guinea. *The New England Journal of Medicine,* **257**, No. 20, 974–8.

16 Hadlow, W.J. (1959). Scrapie and kuru. *The Lancet,* **ii**, 289–90.

17 Gajdusek, D.C., Gibbs, C.J., and Alpers, M. (1966). Experimental transmission of a kuru-like syndrome to chimpanzees. *Nature,* **209**, 794–6.

18 (1998). *New variant of Creutzfeldt-Jakob disease.* Chief Medical Officers Update, 18.

19 Antonowicz, A. (25 January 1994). Give me back my life. *Daily Mirror,* p.1.

20 Buncombe, A. (22 August 1997). Vegetarian girl dying from CJD. *Daily Mail*, p.1.

21 Skegg, D.C.G. (1997). Epidemic or false alarm? *Nature*, **385**, 200.

22 Cousens, S.N., Vynnycky, E., Zeidler, M., Will, R.G., and Smith, P.G. (1997). Predicting the CJD epidemic in humans. *Nature*, **385**, 197–8.

Further reading

Anderson, I. and Nowak, R. (1997). Australia's giant lab. *New Scientist*, **2070**, 34–7.

Butler, D., Wadman, M., Lehrman, S., and Schiermeier, Q. (1998). Last chance to stop and think on risks of xenotransplants. *Nature*, **391**, 320–4.

Collee, J.G. and Bradley, R. (1997). BSE: a decade on—part I. *The Lancet*, **349**, 636–41.

Collee, J.G. and Bradley, R. (1997). BSE: a decade on—part II. *The Lancet*, **349**, 715–21.

Dolan, K., Wodak, A., and Penny, R. (1995). AIDS behind bars: preventing HIV spread among incarcerated drug injectors. *AIDS*, **9**, No. 8, 825–32.

Garred, P. (1998). Chemokine-receptor polymorphisms: clarity or confusion for HIV-1 prognosis? *The Lancet*, **351**, 2–3.

Garrett, L. (1995). *The coming plague. Newly emerging diseases in a world out of balance.* Penguin Books, London.

Hooper, E. (1999). The river: a journey back to the source of HIV and AIDS. Penguin.

Morse, S.S. (1995). Factors in the emergence of infectious diseases. *Emerging Infectious Diseases*, **1**, No. 1, 7–15.

Morse, S.S. (1996). Patterns and predictability in emerging infections. *Hospital Practice*, April 15, 85–104.

Osterhaus, A., Groen, J., Niesters, H., van de Bilt, M., Martina, B., Vedder, L., *et al.* (1997). Morbillivirus in monk seal mass mortality. *Nature*, **388**, 838–9.

Patience, C., Takeuchi, Y., and Weiss, R.A. (1997). Infection of human cells by an endogenous retrovirus of pigs. *Nature Medicine*, **3**, No. 3, 282–6.

Ryan, F. (1998). *Virus X. Understanding the real threat of the new pandemic plagues.* Harper Collins, London.

Wise, J. (1997). HIV epidemic is far worse than thought. *British Medical Journal*, **315**, 1486.

Zhu, T., Korber, B.T., Nahmias, A.J., Hooper, E., Sharp, P.M., and Ho, D.D. (1998). An African HIV-1 sequence from 1959 and implications for the origin of the epidemic. *Nature*, **391**, 594–5.

Chapter 3: Coughs and sneezes spread diseases

1   Bridges, E.L. (1948). *Uttermost part of the earth.* Hodder and Stoughton, London.

2   Creighton, C. (1965). *History of epidemics in Britain.* Vol. 2, (2nd edn), p.308. Frank Cass, London.

3   Schafer Von W. (1955). Vergleichende sero-immunologische untersuchungen uber die viren der influenza and klassischen geflugelpest. *Z. Naturfoschg*, **10b**, 81–91.

4   Taubenberger, J.K. *et al.* (1997). Initial genetic characterization of the 1918 'Spanish' influenza virus. *Science*, **275**, 1793–96.

5   Dr Maria Zambon, Public Health Laboratory Service, Enteric and Respiratory Virus Laboratory, Central Public Health

Laboratory, 61 Colindale Avenue, London NW9 5HT. *Personal communication.*

6   Andrewes, C., Sir. (1973). *In pursuit of the common cold.* William Heinemann Medical Books Limited, London.

Further reading

Burnet, F.M. (1988) Influenza virus A. In: *Portraits of Viruses. A History of Virology* (ed F. Fenner and A. Gibbs) pp. 24–37. Basel, Karger.

Chakraverty, P. (1994). Surveillance of influenza in the United Kingdom. *European Journal of Epidemiology,* **10**, 493–5.

Claas, E.C.J. *et al.* (1998). Human influenza A H5N1 virus related to a highly pathogenic avian influenza virus. *The Lancet,* **351**, 472–3.

Ewald, P.W. (1994). *Evolution of infectious diseases,* pp. 110–18. Oxford University Press, Oxford.

Melnick, J.L. (1988). The Picornaviruses. In: *Portraits of viruses. A history of virology* (ed. F. Fenner and A. Gibbs), pp. 147–88. Basel, Karger.

Waterson, A.P. and Wilkinson, L. (1978). *An introduction to the history of virology.* Cambridge University Press, Cambridge.

Yuen, K.Y. *et al.* (1998). Clinical features and rapid viral diagnosis of human disease associated with avian influenza A H5N1 virus. *The Lancet,* **351**, 467.

Chapter 4: Unlike love, herpes is forever

1   Barré-Sinoussi, F., Chermann, J.C., Rey, F., *et al.* (1983). Isolation of a T-Lymphotropic retrovirus from a patient at risk for acquired immune deficiency syndrome (AIDS). *Science,* **220**, 868–71.

2 Gallo, R.C., Salahuddin, S.Z., Popovic, M., *et al.* (1984). Human T-lymphotropic retrovirus, HTLV-III isolated from AIDS patients and donors at risk for AIDS. *Science,* **224**, 500–3.

3 Crewdson, J. (19 November 1989). The great AIDS quest—science under the microscope. *Chicago Tribune,* Section 5, pp. 1–16.

4 Greenberg, D.S. (1992). The Gallo case: is this really happening? *The Lancet,* **339**, 1594–5.

5 Crewdson, J. (9 February 1992). Inquiry concludes data in AIDS article falsified. *Chicago Tribune,* pp. 1–16.

6 Ho, D.D., Neumann, A.U., Perelson, A.S., *et al.* (1995). Rapid turnover of plasma virions and CD4 lymphocytes in HIV-1 infection. *Nature,* **373**, 123–6.

7 Duesberg, P. and Ellison, B.J. (1996). *Inventing the AIDS virus.* Regnery Publishing Inc., Washington, USA.

8 Weiss, R.A. (1996). AIDS and the myths of denial. *Science and Public Affairs,* pp. 40–44.

9 Darby, S.C., Ewart, D.W., Giangrande, P.L.F., *et al.* (1995). Mortality before and after HIV infection in the complete UK population of haemophiliacs. *Nature,* **377**, 79–82.

10 Easterbrook, P. and Ippolito, G. (1997). Prophylaxis after occupational exposure to HIV. *British Medical Journal,* **315**, 557–8.

11 Johnson, M. (1997). *Working on a miracle.* Bantam Books, New York, Toronto, London, Sydney, Auckland.

12 Zhao, Z–S., Granucci, F., Yeh, L., *et al.* (1998). Molecular mimicry by herpes simplex virus-type 1: autoimmune disease after viral infection. *Science,* **279**, 1344–5.

13 Richmond, C. (1989). Myalgic encephalomyelitis, Princess Aurora, and the wondering womb. *British Medical Journal,* **298**, 1295–6.

14 Report of a Joint Working Group of the Royal Colleges of Physicians, Psychiatrists, and General Practitioners. (1996, revised 1997). *Chronic fatigue syndrome.* Cathedral Print Services Limited, Salisbury.

Further reading

Fukuda, K., Straus, S.E., Hickie, I., *et al.* (1994). The chronic fatigue syndrome: a comprehensive approach to its definition and study. *Annals of Internal Medicine,* **121**, No. 12, 953–9.

Miller, E., Marshall, R., and Vurdien, J. (1993). Epidemiology, outcome and control of varicella-zoster infection. *Reviews in Medical Microbiology,* **4**, 222–30.

Ploegh, H.L. (1998). Viral strategies of immune evasion. *Science,* **280**, 248–53.

Posavad, C.M., Koelle, D.M., and Corey, L. (1998). Tipping the scales of herpes simplex virus reactivation: the important responses are local. *Nature Medicine,* **4**, No. 4, 381–2.

Prusiner, S.B. (1997). Prion diseases and the BSE crisis. *Science,* **278**, 245–51.

Sinclair, J. and Sissons, P. (1996). Latent and persistent infections of monocytes and macrophages. *Intervirology,* **39**, 293–301.

Souberbielle, B.E., Szawlowski, P.W.S., and Russell, W.C. (1995). Is there a case for a virus aetiology in multiple sclerosis? *Scottish Medical Journal,* **40**, 55–62.

Weiss, R.A. and Jaffe, H.W. (1990). Duesberg, HIV and AIDS. *Nature,* **345**, pp. 659–700.

zur Hausen, H. and de Villiers, E–M. (1994). Human papillomaviruses. *Annual Review of Microbiology,* **48**, 427–47.

Chapter 5: Viruses and cancer

1   Burkitt, D. (1962). Determining the climatic limitations of a children's cancer common in Africa. *British Medical Journal,* **2**, 1019–23.

2   Epstein, M.A., Barr, Y.M., and Achong, B.G. (1964). Virus particles in cultured lymphoblasts from Burkitt's lymphoma. *The Lancet,* **(i)**, 702–3.

3   M.A. Epstein, Nuffield Department of Clinical Medicine, John Radcliffe Hospital, Headington, Oxford, OX3 9DU. *Personal communication.*

4   Blumberg, B.S. (1977). Australia antigen and the biology of Hepatitis B. *Science,* **197**, No. 4298, 17–25.

5   Choo, Q–L, Kuo, G., Amy, J., *et al.* (1989). Isolation of a DNA clone derived from a blood-borne non-A, non-B viral Hepatitis genome. *Science,* **244**, 359–61.

6   Poiesz, B.J., Ruscetti, F.W., Gazdar, A.F., *et al.* (1980). Detection and isolation of type C retrovirus particles from fresh and cultured lymphocytes of a patient with cutaneous T-cell lymphoma. *Proceedings of the National Academy of Sciences,* **77**, No. 12, 7415–19.

7   Chang, Y., Cesarman, E., Pessin, M.S., (1994). Identification of Herpesvirus-like DNA sequences in AIDS-associated Kaposi's sarcoma. *Science,* **266**, 1865–9.

8 Kerr, J.F.R., Wyllie, A.H., Currie, A.R. (1972). Apoptosis: a basic biological phenomenon with wide-ranging implications in tissue genetics. *British Journal of Cancer,* **26**, 239–57.

9 Churchill, A.E. and Biggs, P.M. (1967). Agent of Marek's disease in tissue culture. *Nature,* **215**, 528–30.

10 Biggs, P.M. *Personal communication.*

11 Muraskin, W. (1995). *The war against hepatitis B. A history of the international task force on hepatitis B immunization.* University of Pennsylvania Press, Philadelphia, USA.

12 Nimako, M., Fiander, A.N., Wilkinson, G.W.G., *et al.* (1997). Human papillomavirus-specific cytotoxic T lymphocytes in patients with cervical intraepithelial neoplasia grade III. *Cancer Research,* **57**, 4855–61.

13 Rooney, C.M., Smith, C.A., Ng, C.Y.C., *et al.* (1995). Use of gene-modified virus-specific T lymphocytes to control Epstein–Barr virus-related lymphoproliferation. *The Lancet,* **345**, 9–12.

Further reading

Biggs, P.M., Churchill, A.E., Rootes, D.G., and Chubb, R.C. (1968). The etiology of Marek's disease—an oncogenic herpes-type virus. In: *Perspectives in virology,* Vol. 6, pp. 211–37. Academic Press Inc. New York

Blumberg, B.S. (1997). Hepatitis B virus, the vaccine, and the control of primary cancer of the liver. *Proceedings of the National Academy of Sciences (USA),* **94**, 7121–5.

Cancer Research Campaign. (1995). *Cancer—world perspectives.* Factsheet **22.1–22.6.**

Cancer Research Campaign. (1996). *Virus and cancer.* Factsheet **25.1–25.7.**

di Bisceglie, A.M. (1998). Hepatitis C. *The Lancet,* **351**, 351–5.

Duke, R.C., Ojcius, D.M., Young, J.D–E. (1996). Cell suicide in health and disease. *Scientific American,* **December**, 48–55.

Epstein, M.A. (1985). Historical background: Burkitt's lymphoma and Epstein–Barr virus. In: *Burkitt's lymphoma: a human cancer model* (ed. G. Lenoir, G. O'Connor, and C.L.M. Olweny) pp. 17–27, International Agency for Research on Cancer, Lyon. (Scientific Publications, No. 60)

Gallo, R. (1991). *Virus hunting, AIDS, cancer and other human retroviruses.* Harper Collins, London.

Harris, C.C. (1990). Hepatocellular carcinogenesis: recent advances and speculations. *Cancer Cells,* **2**, No. 5, 146–8.

Teodoro, J.G. and Branton, P.E. (1997). Regulation of apoptosis by viral gene products. *Journal of Virology,* **71**, No. 3, 1739–46.

Waterson, A.P. and Wilkinson, L. (1978). *An introduction to the history of virology.* Cambridge University Press, Cambridge.

Wolpert, L. and Richards, A. (1988). Between the lines. In: *A passion in science,* pp. 155–67. Oxford University Press, Oxford.

Chapter 6: Searching for a cure

1   Blower, S.M., Porco, T.C., and Darby, G. (1998). Predicting and preventing the emergence of antiviral drug resistance in HSV-2. *Nature Medicine,* **4**, 673–8.

2   I. William (1998). Department of Sexually Transmitted Diseases, The Mortimer Market Centre, Capper Street, London, WC1E 6AU. *Personal communication.*

3   Lurie, P. and Wolfe, S.M. (1997). Unethical trials of interventions to reduce perinatal transmission of the human immunodefiency virus in developing countries. *New England Journal of Medicine,* **337**, 853–6.

4   Halsband, R. (1953). New light on Lady Mary Wortley Montagu's contribution to inoculation. *Journal of the History of Medicine and Allied Sciences,* **8**, No. 4, 390–405.

5   Fenner, F., Henderson, D.A., Arita, I., *et al.* (1988). Early efforts at control: varioloation, vaccination, and isolation and quarantine. In: *Smallpox and its eradication,* p.260. World Health Organisation, Geneva, Switzerland.

6   Waterson, A.P. and Wilkinson, L. (1978). Early terminology and underlying ideas. In: *An introduction to the history of virology,* p.6. Cambridge University Press, Cambridge.

7   McNeill, W.H. (1976). The ecological impact of medical science and organisation since 1700. In: *Plagues and Peoples,* p.231. Penguin Books, London.

8   P.D. Minor, Division of Virology, National Institute for Biological Standards and Controls, Blanche Lane, South Mimms, Herts, EN6 3QG. *Personal communication.*

9   Minor, P.D., John, A., Ferguson, M., *et al.* (1986). Antigenic and molecular evolution of the vaccine strain of type 3 poliovirus during the period of excretion by a primary vaccinee. *Journal of General Virology,* **67**, 693–706.

10 Wakefield, A.J., Murch, S.H., Anthony, A., *et al.* (1998). Ileal-lymphoid-nodular hyperplasia, non-specific colitis, and pervasive developmental disorder in children. *The Lancet,* **351**, 637–41.

11 Wadman, M. (1997). Clinton sketches out his 'ethical guideposts' for modern biology. *Nature*, **387**, 323.

12 McCarthy, M. (1997). AIDS doctors push for live-virus vaccine trials. *The Lancet*, **350**, 1082.

13 Bower, H. (29 November 1998). New hope. *The Independent on Sunday.*

14 Abrams, D.I. (1999). Getting on with AIDS. *The Lancet*, **353**, 415.

15 (1979). Laboratory work on smallpox virus. *The Lancet*, (**i**), 83–4.

Further reading

Calman, K. (1998). Measles, Mumps, Rubella (MMR) vaccine, Crohn's disease and autism. Department of Health. (PL/CMO/98/2, pp. 1–8.)

Carbone, M., Rizzo, P., and Pass, H.I. (1997). Simian virus 40, polio vaccines and human tumours: a review of recent developments. *Oncogene*, **15**, 1877–88.

Chen, R.T. and DeStefano, F. (1998). Vaccine adverse events: casual or coincidental? *The Lancet*, **351**, 611–12.

Connor, E.M., Sperling, R.S., Gelber, R., et al. (1994). Reduction of maternal-infant transmission of human immunodeficiency virus type 1 with zidovudine treatment. *The New England Journal of Medicine*, **331**, 1173–80.

De Cock, K.M. (1997). Guidelines for managing HIV infection. *British Medical Journal*, **315**, 1–2.

Gambia Government/Medical Research Council Joint Ethical Committee. (1998). Ethical issues facing medical research in developing countries. *The Lancet*, **351**, 286–7.

Graham, B.S., McElrath, M.J., Connor, R.I., et al. (1998). Analysis of intercurrent human immunodeficiency virus type 1 infections in phase I and II trials of candidate AIDS vaccines. *The Journal of Infectious Diseases*, **177**, 310–19.

Kaleeba, N., Ray, S., and Willmore, B. (1991). We miss you all. Noerine Kaleeba: *AIDS in the family*. Women and AIDS Support Network (WASN), Zimbabwe. Printed at Marianum Press.

Wehrwein, P. and Morris, K. (1998). HIV-1-vaccine-trial go-ahead reawakens ethics debate. *The Lancet*, **351**, 1789.

(1998). Vaccine supplement. *Nature Medicine*, **4**, No. 5.

Conclusion: The future—friend or foe?

1   Ryan, F. (1998). *Virus X—understanding the real threat of the new pandemic plagues*. Harper Collins Publishers, London.

2   *Plague wars*, written and reported by Tom Margold. Produced and directed by Peter Molloy. Made by Paladin for BBC and WGBH Frontline.

3   Raina, A.K. and Adams, J.R. (1995). Gonad-specific virus of corn earworm. *Nature*, **374**, 770.

4   Herzog, R.W. *et al.* (1999). Long-term correction of canine hemophilia B by gene transfer of blood coagulation factor IX mediated by adeno-associated viral vector. *Nature Medicine*, **5**, No. 1, 56.

5   Snyder, R.O. *et al.* (1999). Correction of hemophilia B in canine and murine models using recombinant adeno-associated viral vectors. *Nature Medicine*, **5**, No. 1, 65.

6   Rampling, R. *et al.* (1998). Therapeutic replication-competent herpes virus. *Nature Medicine*, **4**, No. 2, 133.

Glossary of terms

Abscess: a localized infection, usually caused by a bacterium. Results in a collection of pus and is associated with tissue damage.

Acquired Immunodeficiency Syndrome (AIDS): an immune deficiency caused by infection with human immunodeficiency virus characterised by opportunistic infections.

Acyclovir: a drug which inhibits the growth of certain herpes viruses. Used mostly to treat or prevent genital and oral herpes and shingles.

Adenovirus: DNA virus which takes its name from the adenoid—the human tissue in which it was first found. Causes respiratory tract and eye infections.

Amino acid: a simple organic compound naturally occurring in plant and animal tissues and forming the basic constituent of proteins.

Anthrax: a disease caused by infection with *Bacillus anthracis* from infected animals. Causes haemorrhage and effusions in various organs and is often fatal.

Antibiotic: a substance which can inhibit or destroy susceptible micro-organisms. (From 'anti-bios', Greek for 'against life').

Antibody: a blood protein capable of binding to and neutralizing an antigen.

Antigen: a foreign substance, usually a protein constituent of a microbe, capable of inducing an immune response in the body.

Antigenic drift: accumulated mutations in flu virus genetic material which eventually differs from the parent virus enough to cause an epidemic.

Antigenic shift: a major genetic change in flu virus arising by genetic reasssortment which may cause a flu pandemic.

Aphrodisiac: a drug which arouses sexual desire. (Named after Aphrodite, the Greek goddess of love.)

Apoptosis: controlled death of cells. (From the Greek 'apo' and 'ptosis', meaning 'falling off'.)

Arborvirus: a large group of RNA viruses, which are spread by arthropod vectors. (Name derived from <u>ar</u>thropod-<u>bor</u>ne viruses).

Aspergillus flavis: *Aspergillis* is a group of fungi including the common moulds. Several are capable of invading the lungs and causing disease. *A. flavis* is a non-pathogenic member of the group which produces penicillin.

Attenuate: reduce in virulence. For a virus, this is usually by prolonged culture under unfavourable conditions.

Autoimmunity: a condition caused by antibodies and/or reactive lymphocytes directed against normal body substances. Causes autoimmune disease.

Bacterium: a unicellular micro-organism.

Boil: a pus-filled swelling caused by infection of a hair follicle; usually with staphylococcus.

Borna disease virus: an unclassified virus which can cause encephalitis in vertebrates including horses, sheep, cats and birds.

Bovine spongiform encephalitis (BSE): a chronic, infectious, degenerative brain disease of cattle, probably caused by prions.

Calcivirus: a RNA virus named after the cup-like depressions seen on the particle surface by electron microscopy.

Candida albicans: *Candida* is a genus of yeast-like fungus. *C. albicans* causes thrush of the mouth and vagina.

CD4: a marker denoting a 'helper' T lymphocyte. The type of cell susceptible to HIV infection.

Cell cycle: the reproductive process of a cell resulting in cell division.

Central nervous system: the nervous tissue of the brain and spinal cord.

Cervix: the neck of the womb or uterus. (From the Latin 'cervix' meaning 'neck'.)

Cholera: an acute epidemic infectious disease caused by the bacterium *Vibrio cholerae,* now occurring primarily in Asia.

Chronic Fatigue Syndrome: excessive, disabling fatigue for at least 6 months duration in the absence of other predisposing conditions.

Cirrhosis: a chronic disease of the liver in which glandular tissue is replaced by fibrous scar tissue. May be caused by alcoholism or hepatitis B or C viruses. (Named by Laennec from the Greek 'Kirrhos' meaning 'tawny'.)

Cohort: a group of people with a common statistical characteristic.

Colitis: inflammation of the colon.

Congenital: existing from birth.

Cornea: clear layer of tissue in front of the eye through which light passes.

Coxackie virus: an enterovirus with two groups (A and B) and many subgroups. Causes respiratory illness, hand, foot, and mouth diseases, and can infect the central nervous system. (Named after Coxackie village in New York State, USA, where the virus was first isolated from two patients with polio-like paralysis.)

Cryptococcus: a yeast found in pigeon faeces which may infect immunocompromised patients.

Cytomegalovirus (CMV): a persistent herpes virus. Infection in utero may cause cytomegalic disease of the newborn, and infection later in life can cause a glandular fever-like illness. In the immunocompromised host, CMV can cause pneumonitis, colitis, and retinitis leading to blindness. (Named after the swollen appearance of infected cells: cyto-megalo = large cell.)

Cytoplasm: the contents of a cell confined by the cell wall and surrounding the nucleus.

Dendritic cell: A cell with processes found in the tissues. Part of the immune system. Dendritic cells trap foreign antigens and carry them to lymph glands to invoke an immune response.

Dengue fever: a flu-like illness consisting of fever, joint and muscle pains, and rash. Caused by the arbovirus, dengue fever virus. Spread by the mosquito *Aëdes aegypti*. The virus can also cause the more severe dengue haemorrhagic fever. (The name is derived from Swahili 'denga' and assimilated into West Indian Spanish meaning 'fastidiousness' referring to the stiffness of a patient's neck and shoulders.)

Distemper: viral disease of animals, especially dogs, causing fever and cough. (From the Latin 'distemperer' meaning 'to derange'.)

DNA: deoxyribonucleic acid. The carrier of genetic information and the constituent of chromosomes. Present in nearly all living cells.

Ebola virus: a filovirus which causes severe and often fatal haemorrhagic fever. Epidemic disease occurs in equatorial Africa. (Named after the river in Zaire where the first recorded outbreak of Ebola fever happened.)

Enterocytopathic human orphan (ECHO) virus: a picornavirus of the genus enterovirus which can cause febrile illness, conjunctivitis, and severe disease in the newborn. (The name refers to the fact that when it was isolated it was not associated with any known disease.)

Electron microscope: a microscope which uses a beam of electrons instead of light. Magnifies up to 100 000 times.

Encephalitis: inflammation of the brain.

Encepthalomyelitis: inflammation of the brain and spinal cord.

Endemic: present in a community or among a group of people. An endemic disease is present continually in a particular region.

Endoplasmic reticulum: a series of membranes inside a cell which form a pathway for proteins produced by the cell to be secreted.

Entamoeba: a family of amoeba, one of which causes amoebic dysentery in man.

Enteroviruses: a genus of the picornavirus family containing viruses which occur in and infect the gut, but do not necessarily cause gastroenteritis. Polio virus is the best-known example.

Enzymes: proteins produced by living cells which catalyse specific biochemical reactions.

Epidemic: an unusual increase in the number of cases of a disease in a community.

Epidemiology: the study of diseaseand disease attributes in defined populations. Epidemiology is the scientific basis for public health and preventive medicine.

Epstein–Barr virus: a herpes virus which is the cause of glandular fever and is associated with a variety of human tumours. (Named after the two discoverers of the virus in 1964.)

Escherichia coli: bacterium that occurs normally in the intestines of man and other vertebrates, is widely distributed in nature, and is a frequent cause of infections. *E. Coli* 0157 has been responsible for recent food poisoning outbreaks.

Filovirus: a family of viruses including Marburg and Ebola viruses. (Name meaning 'thread-like' describes the shape of the viruses—see figure 1.3e).

Flavivirus: a family of arborviruses which includes the yellow fever virus. (Name derived from the Latin *flavus* meaning yellow).

Foot and mouth disease: an infectious disease of cloven-footed animals. Caused by an enterovirus and characterized by a rash, especially in the mouth and in the clefts of the feet.

Gene reassortment: gene swapping in flu viruses which causes an antigenic shift.

Genome: the full set of chromosomes of a cell or organism.

Giardia lamblia: a protozoa which frequently contaminates water supplies. Causes outbreaks of gastroenteritis.

gp120: an HIV-coded protein of molecular weight 120 kilodaltons. Acts as the viral receptor by binding to the CD4 molecule on susceptible cells.

Grey matter: the grey-coloured cellular part of the brain and spinal cord.

Growth factors: soluble proteins, produced by cells, which stimulate the growth of other cells.

Haemagglutinin: a molecule on the surface of flu virus which binds red blood cells.

Haemophilia: an inherited lack of blood clotting factor VIII which leads to a tendency to excessive bleeding when a blood vessel is injured.

Haemorrhagic fever: a syndrome caused by a variety of viruses, including yellow fever, dengue, and Ebola viruses. Causes damage to blood vessels and bleeding which is often fatal.

Hanta virus: a virus of the Bunyavirus family (named after the place where the first member of the family was isolated—Bunyamwera) which is transmitted to humans by rodents. Causes a haemorrhagic fever. (Named after the Hantaan River in Korea where the virus was first identified.)

Hepatitis: inflammation of the liver.

Hepatitis A virus: a picorna virus which causes acute viral hepatitis and is transmitted by the faeca–oral route.

Hepatitis B virus: a hepadnavirus which causes serum hepatitis. Infection may lead to chronic hepatitis, cirrhosis or liver cancer. Usually transmitted by injection of infected blood or by use of contaminated needles. (Name derived from hepa (meaning liver)-DNA-virus).

Hepatitis C virus: a flavivirus which is the principal cause of non-A and non-B post-transfusion hepatitis.

Herpes virus: a family of DNA viruses including those causing cold sores, chickenpox, and shingles. (The name herpes is derived from the Greek 'herpeton' meaning 'reptile' and refers to the creeping nature of the lesions—probably shingles.)

Hormones: substances which influence the action of tissues and organs other than those in which they are produced.

Human immunodeficiency virus (HIV): a retrovirus named after the immune deficiency it causes which leads to AIDS.

Huntington's Chorea: an inherited progressive degenerative disorder of the central nervous system.

Hypoglycaemia: low blood sugar level.

Immunology: the scientific study of immunity, that is, resistance to infection.

Influenza virus. an Orthomyxovirus which causes flu epidemics and pandemics.

JC virus: a polyomavirus which causes a fatal degeneration of the brain—progressive multifocal leucoencephalopathy. Named after the initials of the patient from whom it was first isolated.

Junin virus: an arenavirus which causes Argentinean haemorrhagic fever. (Named after the place in Argentina where the virus was first isolated in 1958.)

Kaposi's sarcoma: a tumour of the cells which line blood vessels (endothelial cells). Common in AIDS patients. Causes red patches on the skin. Is associated with infection with human herpes virus 8.

Kuru: a degenerative brain disease, previously common in the Fore-speaking tribes of Papua New Guinea. (The name in the Fore language means 'shivering' or 'trembling' and denotes the tremor which is characteristic of the disease.)

Lassa fever virus: an arenavirus which causes Lassa fever. (Named after the village in Nigeria, West Africa, where an outbreak occurred which resulted in the isolation of the virus.)

Leukaemia: a malignant disease of the white corpuscles in the blood.

Lymph glands/nodes: a small mass of tissue where lymphocytes congregate and multiply.

Lymphocyte-B: the type of white blood cell which produces antibody.

Lymphocyte-T: the type of white blood cell mainly responsible for immunity to viruses. Includes CD4 'helper' T cells and CD8 'killer' T cells.

Lymphoma: a tumour of lymphoid tissue.

Lyssavirus: a genus of viruses including the rabies virus. (Name derived from the Greek *Lyssa* meaning madness).

Macrophage: a bone marrow-derived cell which engulfs and digests foreign or dead material in tissues.

Marek's disease virus: a herpesvirus which causes tumours in chickens.

Morbillivirus: the genus of the paramyxovirus family to which measles virus belongs. ('Morbilli' is a Latin diminutive of 'morbus' meaning 'a disease', denoting a minor illness.)

Motaba River Valley virus: a fictitious virus used in the film *Outbreak*.

Mucous membrane: the general name given to the membrane which lines many of the hollow organs of the body.

Multiple sclerosis (MS): a disease of the brain and spinal cord. Over a period of time it can cause paralysis and tremors.

Mutation: an instant genetic change or alteration which, when transmitted to offspring, gives rise to inheritable variations.

Myalgia: pain in a muscle or group of muscles.

Myalgia encephalomyelitis: an alternative name for chronic fatigue syndrome.

Myelin basic protein: a major protein constituent of nerve tissue.

Myxomatosis: a fatal disease of European rabbits marked by conjunctivitis and the development of myxomatous growths in the skin; caused by rabbit myxoma virus.

Neuraminidase: an enzyme on the surface of flu viruses which can destroy neuraminic (sialic) acid.

Nucleus: the central body in a cell which contains the chromosomes.

Oncogene: a gene which can transform a cell into a tumour cell.

Pandemic: an epidemic involving more than one continent at once.

Papillomaviruses: a genus of viruses which cause benign warts and malignant epithelial tumours of uterine cervix, penis, and larynx. (Named from Latin 'papilla' meaning 'nipple'.)

Parainfluenza viruses: a family of viruses which cause acute respiratory illnesses including croup.

Paramyxovirus: a family of viruses including those which cause measles, mumps, and respiratory tract infections such as croup in babies.

Parkinson's disease: a progressive disease of the nervous system with tremor, muscular rigidity, and emaciation.

Parvovirus: a family of viruses including the cause of the childhood illness erythema infectiosum (infectious rash or fifth disease). (Named from the Latin 'parvus' meaning 'small'.)

Pathogen: any micro-organism causing disease.

Penicillium notatum: an antibacterial substance produced by a mould. (Sir Alexander Fleming discovered its antibacterial action in 1929.)

Peptide: a compound formed by the union of two or more amino acids which can itself form part of a protein.

Perinatal: relating to the time immediately before and after birth.

Picornavirus: a virus of the family Picornaviridae. (Name derived from Pico (small), RNA virus).

Placebo: a medicine (or pill) given which will not have any definite action.

Placenta: the afterbirth.

Plague: the name of an infectious epidemic disease caused by the bacterium *Yersinia pestis*. Common in the Middle Ages. Causes fever and swelling of the lymphatic glands and carries a very high mortality. Rats are the natural reservoir.

Plasmid: DNA which is separate from the chromosome of the host cell (chiefly bacterial) and can replicate but is not essential for the cell's survival.

Pneumocystis carinii: a single-celled organism related to fungi. Causes pneumonia in the immunocompromised host.

Pneumonitis: inflammation of the lungs that does not progress to pneumonia.

Polio virus: an enterovirus which causes poliomyelitis. Derived from the Greek 'polios' meaning 'grey' and 'myelos' meaning 'marrow'.

Polyoma virus: a virus of the Papovavirus family which cause tumours in animals. (Name derived from the Greek *Poly* meaning many, *oma* meaning tumours).

Pox viruses: a family of viruses including the cause of smallpox. (Derived from the Anglo-Saxon word 'pokkes' meaning 'pouch' and referring to the skin lesions.)

Primate: the highest order of mammals which includes tarsiers, lemurs, apes, monkeys, and humans.

Prion (proteinaceous infectious particle): an infectious protein believed to be responsible for transmissible neurodegenerative diseases.

Programmed cell death: another term for apoptosis.

Protease: an enzyme which can break down proteins and peptides; a proteolytic enzyme.

Protein: nitrogenous substance widely distributed in the animal and vegetable kingdoms and forming the characteristic materials of tissues and fluids. Essentially combinations of amino acids.

Protozoa: free-living, mobile, unicellular organisms, including amoebae, which can induce dysentery and plasmodium (the cause of malaria).

Rabbit haemorrhagic fever virus: a calcivirus which causes severe haemorrhagic fever in susceptible rabbits.

Rabies: a fatal disease which constantly infects dogs and wolves in some countries. Infection is from the saliva of a rabid animal. Caused by a virus of the rhabdovirus family (derived from the Greek 'rhabdos' meaning 'rod').

Receptor: a molecule in a cell membrane which binds specifically to a substance.

Repetitive stress injury: injury arising from the prolonged used of particular muscles.

Respiratory syncytial virus: one of the myxoviruses. This is the major cause of bronchiolitis and pneumonia in infants under the age of six months.

Retinitis: inflammation of the retina.

Retroviruses: a group of RNA viruses which insert a DNA copy of their genome into host chromosomes in order to persist.

Reverse transcriptase: an enzyme which catalyses the conversion of RNA to DNA. Carried by retroviruses and required for their integration into the host cell genome.

Rhabdovirus: a family of rod-shaped viruses of the lyssavirus genus including the rabies virus. (Name derived from the Greek *rhabdos* meaning a rod).

Rhinovirus: a picornavirus which causes the common cold. (The name is derived from the Greek 'rhis' meaning 'nose'.)

Rindepest virus: virus of the genus *Morbillivirus*, causing the acute, highly contagious disease Rindepest in ruminants and pigs.

RNA: ribonucleic acid. One of the two types of nucleic acid that exist in nature (the other is DNA). It is present in both the cytoplasm and nucleus of cells.

Rotavirus: a group of viruses so-called because of their wheel-like structure. A common cause of gastroenteritis in infants. (Derives from the Latin 'rota' for 'wheel'.)

Rubella: another name for German measles. ('Rubellus' being Latin for 'reddish' and referring to the rash of rubella.) Caused by rubella virus, a togavirus (the name being derived from the toga-like, close-fitting envelope surrounding the virus).

Ruminant: an animal that chews the cud regurgitated from its rumen.

Salmonella: bacteria of the genus *Salmonella*, some of which cause food poisoning.

Scrapie: a disease of sheep involving the central nervous system. Characterized by lack of co-ordination and itching (causing affected animals to rub against trees etc.) Thought to be induced by a prion.

Shigella: the name given to a group of rod-shaped, negative bacteria that are the cause of bacillary dysentery.

Simian immunodeficiency virus (SIV): a retrovirus which causes an AIDS-like syndrome in certain species of old-world primates.

Sin nombre virus: a type of Hanta virus. (Derived from the Spanish for 'without name'.)

Slim disease: a wasting disease commonly seen in African AIDS.

Stem cell: an undifferentiated cell from which specialized cells develop.

Streptococcus: round bacteria which under the microscope have the appearance of a string of beads. Responsible for throat infections and other forms of inflammation.

Stromal herpes keratitis: herpes virus infection of the eye which can cause scarring leading to blindness.

Subunit vaccine: vaccine made from parts of viruses, such as individual proteins, which induce an immune response.

Theiler's virus: a Picornavirus which causes auto-immune disease of the central nervous system in mice.

Thrush: an infection affecting particularly the mouth and causing a patchy white appearance on the lips, tongue, and palate. It is caused by a fungus (*Candida albicans*) growing on the surface of the mucous membrane.

Tissue culture: the growth in an artificial medium of cells derived from living tissue.

Togavirus: a virus family including the cause of rubella (so named because of the closely fitting (toga-like) virus envelope).

Toxin: poison produced by bacteria. Usually soluble and easily destroyed by heat.

Toxoplasma: a protozoa which can cause disease congenitally or in the immunocompromised host.

Transmissible Spongiform Encephalitis: a group of transmissible degenerative diseases of the central nervous system thought to be caused by prions.

Trigeminal nerve: the fifth cranial nerve which supplies the skin of the face. It has three divisions: the ophthalmic, maxillary, and mandibular nerves.

Tropical spastic paraparesis: a chronic neurological disease which occurs in the Caribbean and Japan and is associated with human T cell leukaemia virus infection.

Tuberculosis (TB): the diseases caused by *Mycobacterium tuberculosis*, of which pulmonary tuberculosis is the most important.

Tumour suppressor genes: genes which negatively control cell division. Mutations of these genes are associated with many types of cancer.

Vaccination: the process of inoculating with a vaccine to obtain immunity, or protection, against the corresponding disease. (Derived from the Latin 'vacca' for 'cow'.)

Vaccine: a substance, either dead or attenuated living infectious material, introduced into the body to produce resistance to a disease.

Vaccinia virus: a pox virus used in vaccination against smallpox. The origins of vaccinia virus are obscure.

Varicella: another name for chickenpox.

Viroid: an infectious entity affecting plants. Similar to a virus but smaller and consisting only of nucleic acid without a protein coat.

Virus Particle: the extracellular form of a virus consisting of a protein coat or capsid surrounding the genetic material (DNA or RNA).

Yellow fever: an acute viral infection of certain tropical localities. Characterized by fever and jaundice.

Zidovudine (AZT): a derivative of thymine used to treat HIV and other viral infections.

Zoster: or shingles. A skin eruption, caused by varicella zoster virus, consisting of small yellow vesicles. (From the Greek word 'circingle' meaning 'girdle', because it spreads in a zone-like manner round half the chest.)

Zoonosis: a disease of humans acquired from an animal source.

Index